Submerging Coasts

Coastal Morphology and Research

Series Editor: Eric C. F. Bird

CORAL REEF GEOMORPHOLOGY
André Guilcher

COASTAL DUNES
Form and Process

Edited by
Karl Nordstrom, Norbert Psuty and Bill Carter

GEOMORPHOLOGY OF ROCKY COASTS
Tsuguo Sunamura

SUBMERGING COASTS
The Effects of a Rising Sea Level on Coastal Environments
Eric C. F. Bird

Submerging Coasts

The Effects of a Rising Sea Level on Coastal Environments

Eric C. F. Bird
Geostudies, UK

UNEP

The publication of this book was sponsored by
United Nations Environment Programme, Nairobi, Kenya

JOHN WILEY & SONS
Chichester · New York · Brisbane · Toronto · Singapore

Other Wiley Editorial Offices

John Wiley & Sons, Inc., 605 Third Avenue,
New York, NY 10158-0012, USA

Jacaranda Wiley Ltd, G.P.O. Box 859, Brisbane,
Queensland 4001, Australia

John Wiley & Sons (Canada) Ltd, 22 Worcester Road,
Rexdale, Ontario M9W 1L1, Canada

John Wiley & Sons (SEA) Ptc Ltd, 37 Jalan Pemimpin #05-04,
Block B, Union Industrial Building, Singapore 2057

Library of Congress Cataloging-in-Publication Data

Bird, Eric C. F. (Eric Charles Frederick), *1930–*
    Submerging coasts : the effects of a rising sea level on coastal
  environments / Eric C.F. Bird.
      p.   cm. — (Coastal morphology and research)
    Includes bibliographical references and index.
    ISBN 0-471-93807-6 (pbk.)
    1. Coast changes.   2. Sea level.   I. Title.   II. Series.
  GB451.2.B58   1993
  551.4'58—dc20   92–30175
                                              CIP

British Library Cataloguing in Publication Data

A catalogue record for this book is available from the British Library

ISBN 0-471-93807-6

Typeset in 10/12pt Times from author's disks by Text Processing Department,
John Wiley & Sons Ltd, Chichester
Printed and bound in Great Britain by Biddles Ltd, Guildford, Surrey

Contents

Preface

This book is a sequel to my *Coastline Changes: A Global Review*, which was published in 1985, and initiated the Coastal Morphology and Research series. At about that time, there was a sudden surge of interest in climatic change and sea level variation, stimulated by evidence that the composition of the Earth's atmosphere had been modified by the discharge of increasing amounts of carbon dioxide, methane and nitrous oxide generated by urban, industrial and agricultural activities. These are among the gases that produce the so-called greenhouse effect, which maintains Earth's surface temperature at a much higher level than would otherwise exist: their enhancement by human agency is likely to cause global warming and a world-wide sea level rise.

At present, submerging coasts are largely confined to areas where the land margin is subsiding, so that a relative rise of sea level is in progress. They are also to be found around the Caspian Sea and the Great Lakes in North America during phases of rising water level. The prospect now is that most of the world's coastline will soon be submerging because of global warming and world-wide sea level rise, and there has been much discussion of the coastline changes that will occur, and of how people living in coastal regions will respond to these changes.

The question of the nature and dimensions of climatic change and sea level variation has been dealt with in a number of books and monographs. The present work seeks to determine what is happening on coasts that are already submerging, and to indicate the changes likely to occur as sea level rise develops globally. Attention is given to geomorphological and associated ecological changes, and then to human responses, as influenced by social, economic and political considerations.

The author has been concerned with these problems, especially on behalf of the United Nations Environment Programme (UNEP), which has sponsored research on and assessment of the effects of a rising sea level in various parts of the world. There have been many conferences and training workshops, and the topic has generated a rapidly growing scientific literature, as well as continuing media interest. In 1992 I was asked to prepare a global review for UNEP, which has been used in writing this book.

The aim is to provide a background for discussion and evaluation of problems resulting from coastal submergence, in particular for those concerned with the planning, development, conservation and management of coastal areas. This book should also be of use to people concerned with education and training, and media publicity, when they address problems of a rising sea level and their consequences for coastal environments.

I am grateful to UNEP, notably to Peter Usher, co-ordinator of the Climate Unit, and Peter Schröder of the Regional Seas Programme, for supporting my studies and providing information. I am also indebted to my colleagues in the International Geographical Union's Commission on the Coastal Environment (now the Commission on Coastal Systems) for discussions and information exchange through my project on the physical and human impacts of a rising sea level (1988–1992). My thanks go also to Chandra Jayasuriya and Wendy Nicol of the Department of Geography, University of Melbourne, for their preparation of maps, diagrams and photographs, and to Lilian Modlock of Geostudies (UK) for assistance in processing the text of this book.

Eric C.F. Bird
Falmouth, July 1992

Chapter One

Introduction

The world's coastline is about a million kilometres in length, and consists of a variety of landforms including cliffs, bluffs, beaches, dunes, estuaries, lagoons, deltas, coastal plains, salt marshes and mangroves. There are also fringing and outlying coral reefs and associated low-lying islands. The evolution of these features, and the ways in which they change in response to coastal processes, have been discussed in several textbooks (e.g. Bird 1984, Carter 1988, Davies 1972, Pethick 1984), while the nature and pattern of coastline changes over the past century, documented by the International Geographical Union's Commission on the Coastal Environment (IGU-CCE) between 1972 and 1984, were summarised by Bird (1985). As well as changes due to the erosion, transport and deposition of coastal sediments there have been changes resulting from variations in the level of land and sea around the world's coastline, and changes produced by the building of artificial structures and by land reclamation projects on the coast.

Sea level has remained relatively stable around much of the world's coastline over the past few thousand years, but some sectors have been emerging because sea level has fallen relative to the land, while others are submerging because of a relative sea level rise. Such changes may be due to interactions of land uplift with actual rises or falls in the level of the sea. In northern Scandinavia, for example, coastal land has been emerging from the sea and the coastline has been advancing largely because of uplift, while the gradual submergence taking place along the southern and eastern coasts of England is thought to be due partly to land subsidence and partly to a rising sea.

During the past decade much attention has been given to climatic changes expected as a result of the accumulation in the Earth's atmosphere of such gases as carbon dioxide, nitrous oxide and methane, produced mainly by industry and agriculture. It is thought that the increasing concentrations of these gases will enhance the natural greenhouse effect, whereby the atmosphere intercepts some of the solar radiation reflected into space from the Earth's surface, and so maintains global temperatures at a higher level than would otherwise prevail.

The forecast is therefore a human-induced global warming, and this in turn has led to predictions of a world-wide sea level rise.

During the 1970s, many climatologists held the view that the Earth was about to become much cooler, with the present interglacial giving place to a new glacial phase. An alternative prediction grew out of recognition, from the results of global atmospheric monitoring initiated during the International Geophysical Year in 1957, that concentrations of carbon dioxide and other gases in the atmosphere were indeed steadily increasing, and the realisation that such an increase could enhance the greenhouse effect. Mercer (1978) raised the alarm about the possible disintegration and melting of the West Antarctic ice sheet leading to a sudden rise of global sea level by several metres. In 1982 the United States Environmental Protection Agency (EPA) launched a project on possible future changes in sea level resulting from increasing carbon dioxide in the atmosphere (Revelle 1983), and in April 1983 the EPA held a conference that led to the publication of an influential book, *Greenhouse Effect and Sea Level Rise* (Barth and Titus 1984). Hoffman (1984) then estimated that global sea level could rise between 56.2 and 345.0 cm by the year 2100, and Hansen et al. (1984) reported that mean global air temperature was rising, having already increased 0.4°C during the past century. In 1985 the Villach Conference, held in Austria, issued a statement indicating that mean global air temperature was expected to rise between 1.5°C and 4.5°C during the coming century, and that this would lead to a world-wide sea level rise of between 0.20 and 1.40 m. In June 1986 the US EPA, in association with the United Nations Environment Programme (UNEP), held a meeting in Washington on 'The Effects of Changes in Stratospheric Ozone and Global Climate' (Titus 1986a), and in August of that year the 'Impact of Sea Level Rise on Society' was discussed at a meeting in The Netherlands (Wind 1987). In 1987 the Commonwealth Scientific and Research Organisation (CSIRO), Australia, held a symposium in Melbourne which examined the issues from an Australian perspective in 'Greenhouse: Planning for Climatic Change' (Pearman 1988), and a number of other regional and national conferences followed, together with numerous reviews (e.g. Jones and Henderson-Sellers 1990, Warwick and Farmer 1990). In 1990 the Intergovernmental Panel on Climate Change (IPCC) reviewed the evidence and proposed the scenarios shown in Fig. 1, with high and low estimates, and a 'most likely' intermediate prediction that global sea level will rise by about 20 cm by the year 2030 and accelerate to attain 1 m some time in the 22nd century. The present study examines the changes that would result from a sea level rise of 1 m around the world's coastline.

An obvious outcome will be that submerging coastlines, presently confined mainly to sectors where the land has been subsiding (Fig. 2), will become more extensive, and that emerging coastlines will become rarer. In due course, all of the world's coastline will be submerging as a result of sea level rise. This book

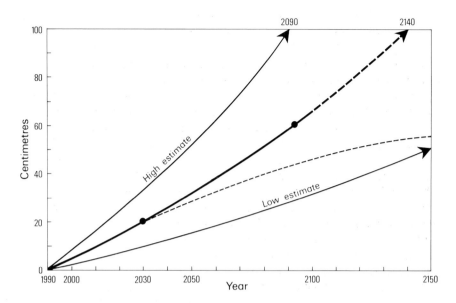

Figure 1   Scenarios of predicted global sea level rise presented by the Intergovernmental Panel on Climate Change (Houghton et al. 1990). The most likely scenario (bold line), a rise of 20 cm by the year 2030, 60 cm by the year 2090, and 1 m by the year 2140, is the one adopted in this study, but high estimates (1 m by 2090) and low estimates (50 cm by 2150) were also offered. If the enhancing of the greenhouse effect were halted by the year 2030, sea level would continue to rise at a slackening rate for more than a century (dashed line)

is concerned with the changes in coastal environments that are likely to result from such a sea level rise, and with possible human responses to these changes.

⚡ Many geomorphologists and oceanographers are of the opinion that a world-wide sea level rise, of the order of 1–2 mm per year, is already in progress, and that the changes predicted as a consequence of enhancing the greenhouse effect have commenced; others consider that this cannot yet be demonstrated. As will be seen, there is a need for caution in predicting the nature and rate of change, both of climate and sea level, on the basis of existing meteorological and oceanographic evidence, but there is wide agreement that global warming and sea level rise will occur during the coming century. If the sea level rise proves to be either more, or less, rapid than the IPCC intermediate scenario (Fig. 1), then adjustments can be made to the rate at which the resulting changes will take place.

⚡ It must be emphasised that changes are already taking place around the world's coastline, and that these would continue even if no sea level rise took place. It becomes a question of deciding how existing changes will be modified by

4

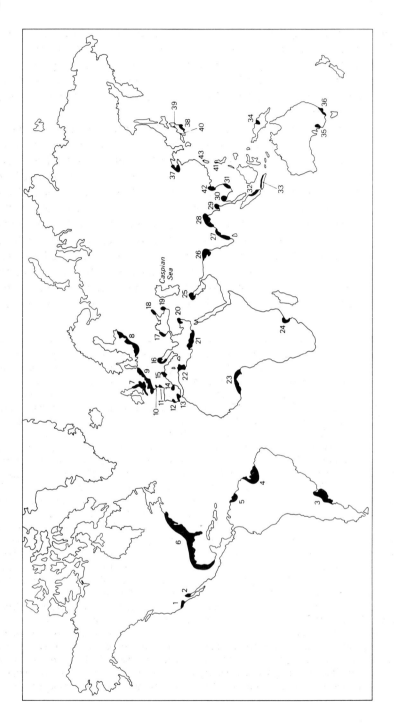

a rising sea level, a question that can be examined on coastal sectors where land subsidence has already produced a relative sea level rise (Milliman 1988), reviewed in terms of geological evidence of changes that accompanied past episodes of sea level rise (marine transgressions) (Thom and Roy 1987), or treated theoretically from known relationships between coastal processes and landform evolution.

Some preliminary definitions are necessary. The coast consists of a number of zones of varying width: the shore (or foreshore) zone between normal high and low tide lines, the nearshore zone to seaward, usually out to the line where waves begin to break, and the offshore zone beyond that, with a backshore zone extending landward from the high tide line. The landward limit of the coast may be defined by landform features, such as the first ridge crest or the watershed of small coastal catchments, or the boundaries of coastal plains, beach ridge and dune systems, lagoons or swamps. Alternatively, a contour 5 or 10 m above sea level may be used. There are problems where the hinterland rises gradually and where coastal lowlands extend inland along valley floors. Coastal managers, planners and administrators sometimes prefer arbitrary boundaries: 1 km or some other chosen distance inland. The coastline is defined as the land

---

Figure 2 Sectors of the world's coastline that have been subsiding in recent decades, as indicated by evidence of tectonic movements, increasing marine flooding, geomorphological and ecological indications, geodetic surveys, and groups of tide gauges recording a rise of mean sea level greater than 2 mm per year over the past three decades.

Key to map: 1, Long Beach area, southern California; 2, Columbia River delta, head of Gulf of California; 3, Gulf of La Plata, Argentina; 4, Amazon delta; 5, Orinoco delta; 6, Gulf and Atlantic coast, Mexico and United States; 7, southern and eastern England; 8, the southern Baltic from Estonia to Poland; 9, northern Germany, The Netherlands, Belgium and northern France; 10, Loire estuary, western France; 11, Vendée, western France; 12, Lisbon region, Portugal; 13, Guadalquavir delta, Spain; 14, Ebro delta, Spain; 15, Rhône delta, France; 16, northern Adriatic from Rimini to Venice and Grado; 17, Danube delta, Romania; 18, eastern Sea of Azov; 19, Poti Swamp, Georgian Black Sea coast; 20, south-east Turkey; 21, Nile delta to Libya; 22, north-east Tunisia; 23, Nigerian coast, especially the Niger delta; 24, Zambezi delta; 25, Tigris–Euphrates delta; 26, Rann of Kutch; 27, south-eastern India; 28, Ganges–Brahamputra delta; 29, Irrawaddy delta; 30, Bangkok coastal region; 31, Mekong delta; 32, eastern Sumatra; 33, northern Java deltaic coast; 34, Sepik delta; 35, Port Adelaide region; 36, Corner Inlet region; 37, Hwang-ho delta; 38, head of Tokyo Bay; 39, Niigata, Japan; 40, Maizuru, Japan; 41, Manila; 42, Red River delta, North Vietnam; 43, northern Taiwan.

The Caspian coast is also indicated because the Caspian Sea, having fallen in level between 1930 and 1977 by about 3 m, has since 1976 risen just over 1.5 m (see Fig. 12, page 23). The fluctuations of level are considered to be due to climatic variations influencing river discharge into the Caspian Sea, but the completion of a dam across the mouth of the Kara Bogaz Gol, Zaliv embayment, an area of high evaporation, may also have contributed. The present submergence of the Caspian coasts is providing a 'field laboratory' for studies of how coastal features are modified by a rising sea level

margin at normal high tide, whereas the shoreline moves to and fro as tides rise and fall: there is a low-tide shoreline, a mid-tide shoreline, and a high-tide shoreline. In American literature, however, the term 'shoreline' is often used as a synonym for 'coastline' and the term 'coast' is elaborated into 'coastal zone'. As sea level rises, the location, geomorphology and associated ecology of each of these coastal zones will change.

The term 'coastal environment' is taken to include the land, air and water components of the coast as thus defined, and it is convenient to refer in a general way to coastal areas and coastal regions.

Chapter 2 deals with the problems of defining present sea level, of detecting and monitoring (i.e. measuring and recording) a sea level rise, and of predicting how far and how fast sea level will change as a result of expected global warming. Chapter 3 then examines the geomorphological and related ecological changes likely to occur as sea level rises in various coastal environments. For each of these, there is a brief consideration of how the existing landforms have been shaped and how they are now changing, before the effects of a rising sea level are discussed. The changes outlined are initially those that would occur in the absence of any human attempt to modify or prevent them, but it is acknowledged that, just as existing coastlines show the effects of human impacts, so will the changes that occur with a rising sea be modified in places by attempts to prevent erosion and efforts to reclaim land from the sea. Parts of the world's coastline are already quite artificial, and the prospect is that these will be maintained and extended, and that the extent of natural coastline will diminish, as attempts are made to counter the effects of a global sea level rise.

Chapter 4 discusses the kinds of response that have taken place on coasts where the sea has been rising because of land subsidence, as a basis for predicting the effects of a global sea level rise. It then explores three possible scenarios: to retreat as the coastline recedes, adapting human activities, where possible, to the changing environment as submergence proceeds; to try to maintain existing coastlines with defensive structures; or to counter-attack by advancing the coastline seaward by means of land reclamation projects. A final section discusses the many social, economic and political factors that will complicate the responses that actually occur as sea level rises around the world's coastline.

There is a rapidly growing literature on global sea level rise, its expected environmental consequences and probable human responses, together with implications for coastal planners, developers, managers and conservationists. The published material is widely scattered, and some of it is difficult to find. There are also many unpublished reports, some of very limited circulation. Inevitably, much of this literature is generalised and speculative, and media coverage has often been unnecessarily sensationalised. On the other hand, few would now accept the attitude expressed by Wanless (1982), who entitled a paper 'Sea level is rising – so what?'. Recent papers that have dealt with the general problems of sea level rise include those by Viles (1989), Tegart, Sheldon and Griffiths (1990) and Stoddart and Reed (1990). A United Nations Educational, Scientific, and Cultural Organization (Unesco) report has critically evaluated the question of sea level changes (Stewart et al. 1990), and there have been a number of regional studies, some focusing on the problems of coastal cities (Frassetto 1991). These include the United Kingdom (Department of the Environment 1991), the United States (Gornitz and Kanciruk 1989), the Mediterranean (Jeftic, Milliman and Sestini 1992, Sestini 1992), the Caribbean (Hendry 1988, Nurse 1992), south-east Asia (Paw and Chua Thia-Eng 1991), and the Pacific region (Pernetta and Hughes 1992).

The topics here discussed include reference to publications that were available to the author by the early months of 1992. There are, as will be seen, many uncertainties, and the aim is to provide a background for further discussion, research and the prediction of environmental changes and human responses on particular sectors of coastline. Sea level is expected to rise gradually at first, and then at an accelerating rate, so the response to coastline changes is likely to be incremental, decade by decade. The response will be stronger after unusual events such as storm surges, major floods, earthquakes and tsunamis focus attention on the problems of a rising sea. As submergence and erosion accelerate, their effects are likely to prompt a long-term strategy for coastal utilisation and development, and the sooner the probable changes are understood the more effective will be the response of planners, developers, managers and conservationists to the problems that ensue. In the long run, mankind will have to reach agreement on the management of the global atmosphere and oceans in such a way as to stabilise the sea at an optimal level around the world's coastline. The United Nations Conference on Environment and Development (UNCED) in Rio de Janeiro in June 1992 was a step in this direction. Meanwhile, this account of submerging coasts, past, present and future, is offered as a contribution to the discussion.

Chapter Two

Changing sea level

INTRODUCTION

The level at which the sea stands, relative to the land, around the world's coastline depends partly on the volume of water in the oceans, which is determined largely by the balance of evaporation and precipitation produced by the hydrological cycle, and partly on the size and shape of the crustal depressions that contain the oceans. In the geological past the sea has sometimes stood at higher levels, indicated by former shore and beach features above present sea level, and sometimes at lower levels, on coastlines that can now be found submerged on the sea floor. These changes of sea level have resulted either from the uplift or lowering of coastal areas (tectonic movements) or the rise and fall of the sea surface (eustatic movements), or some combination of the two.

Sea level is variable over many time-scales. It oscillates with the tides produced by the gravitational effects of the sun and the moon in relation to the Earth. The mean vertical range from high to low tide varies considerably around the world's coastline, from almost zero to more than 20 m. It increases at spring-tides, when the gravitational forces of the sun and the moon are combined, and diminishes at neap-tides, when they are opposed. Variations in tide range over time complicate the determination of present sea level.

Statements about present sea level usually refer to mean sea level, which can be estimated by the long-term average of high and low tide levels, i.e. mean tide level. Strictly, mean sea level is defined as the arithmetic mean of the height of calm sea surface measured at hourly intervals over a period of at least 19 years (see below). In practice, it is defined in relation to some national datum such as American Sea Level Datum, or Ordnance Datum in the United Kingdom (Kidson 1986). Averaging is necessary to eliminate the variations that occur over tidal cycles that range from semi-diurnal (high and low tides recurring at intervals of about 12.8 hours) through diurnal (one high and one low tide every 24 hours) to the 14.6-day spring-tide and neap-tide cycle already mentioned, and longer term oscillations of which the 18.6-year 'nodal tide', based on the precession of the lunar orbit, is best known (Pugh 1987a). The latter reached maxima in 1913.4±18.6 years: there was one in October 1987, and the next is due in May 2006. Mean high and mean low tides rise and fall with these oscillations, which in theory are symmetrical about the long-term sea level, but in practice mean sea level determinations have traditionally spanned at least 19 years. Use has been made of annual and decadal means and running means in attempts to discover whether sea level changes have occurred during the period for which tide gauge records are available. Even longer term astronomical tidal cycles have been identified, with maxima occurring in 1745 and 1922, and another expected in 2192 (Cartwright 1974): by that time they may be superimposed on a rather different mean sea level.

Sea level also fluctuates in relation to weather conditions, notably the variations in atmospheric pressure that accompany the passage of depressions and anticyclones, together with the effects of associated winds. Strong wind action (cyclones, typhoons, hurricanes) can produce storm surges that drive water shoreward and thus raise sea level temporarily on a lee shore by up to several metres, as well as producing very high waves that break upon the coast. Persistent onshore winds can maintain high water levels, especially in gulfs and semi-enclosed bays. Sea level also varies globally in relation to seasonal temperature, pressure and wind regimes. The response to atmospheric pressure changes is usually stated as a sea level rise of about 1 cm for each millibar fall. Analyses of monthly or seasonal data show that maximum annual sea levels occur at different times around the world's coastline: in September in the eastern North Atlantic, for example, and in April along the eastern seaboard of Australia (Guilcher 1965). In the South China Sea mean sea level is about 40 cm higher during the north-east monsoon (November to March) than during the south-west monsoon.

Sea level also varies in relation to atmospheric pressure cycles such as the North Atlantic Oscillation, which is correlated with sunspot cycles, and the 2–7 · year El Niño Southern Oscillation (ENSO), which produces high pressure and low sea level over the south-east Pacific and low pressure and high sea level in the Indian Ocean, then reverses these (Fairbridge and Krebs 1962). Komar (1986) showed that the 1982–1983 ENSO raised sea level 20–30 cm along the Pacific coast of North America. Such variations complicate the oscillations registered on tide gauges, and should be excluded from the record when present mean sea level is determined.

Determination of sea level, and of sea level changes, depends heavily on data from tide gauges, and on the accuracy of tide gauge records. Most tide gauges are instruments that were installed for the purposes of navigation and harbour procedures, rather than to provide scientific information for a global assessment. Many are located on port structures (Fig. 3), where there may be

Figure 3   The tide gauge at Newlyn Harbour, in Cornwall (south-west England) is located in a building (arrowed) on the southern quay, near the lighthouse. It was used to establish mean sea level for the period 1915–1921 as Ordnance Datum for British topographical surveys, but by the 1960s mean sea level had risen 0.12 m here, an annual average rise of 2.4 mm per year

local tidal anomalies resulting from wave reflection and ponding, and where they are exposed to damage and disturbance in the course of ship movements and port operations (Fig. 4). They register actual fluctuations of the sea surface, including the effects of wind stress, waves, and changes in atmospheric pressure in relation to datum levels that have sometimes been poorly maintained. They are subject to local effects, such as current swirl, especially in harbour areas; and if a disturbance or accident results in subsidence of their foundations, this will be recorded as a spurious rise in mean sea level. More scientific tide gauge instrumentation is being developed around the world's coastline, and this will provide improved monitoring of sea level changes in the future. Some caution is advisable in the interpretation of particular tide gauge data, especially if local anomalies are indicated.

An alternative source of information on sea level changes in recent decades has been provided by repeated geodetic surveys of land areas, based on precise levelling. In Sweden and Finland, for example, successive geodetic surveys have confirmed the fall in sea level indicated by tide gauges along the shores of the Bothnian Gulf, which results from continuing land uplift in Scandinavia

Figure 4   The tide gauge at Port Adelaide, South Australia, is housed in a box on a wooden pier. Records from this tide gauge indicated an annual average rise of mean sea level of 2.5 mm per year between 1882 and 1976

following deglaciation. Evidence of transverse tilting of the British Isles, the north and west having risen while the south and east subsided, was deduced by Valentin (1953) from analyses of tide gauge records, but no firm conclusions could be drawn from comparisons of geodetic surveys of the British Isles in 1840–1860 and 1912–1921 because of errors in the first levelling, and doubts on the accuracy of the second have complicated interpretation of a third in 1951–1956 (Kidson and Heyworth 1979). Woodworth (1987) also found evidence of the transverse tilting from further tide gauge analyses, which showed mean sea level rising faster at Sheerness and Newlyn than at North Shields and Aberdeen, but geodetic confirmation is still awaited. Geodetic surveys are elaborate and expensive, and it will be some time before they can contribute much to knowledge of global sea level changes. Monitoring from satellites will in due course provide an integrated global tidal facility, with more accurate information on the rise and fall of land and sea levels around the world, using measurements based on the Earth's centre as a datum point. It will then be possible to distinguish the contributions of land uplift or subsidence and sea surface rise or fall to the vertical changes in sea level relative to the land registered in tide gauge records (Goldsmith and Hieber 1991). This is for the future. For the detection of past and present changes we have to rely on results obtained from tide gauge records.

Although much attention has been given to determining mean sea level, and discovering whether it has changed, the levels reached by high tides are of more importance to people living in coastal areas. When exceptionally high tides occur they prompt questions about possible changes in sea level. In the Venice region, for example, where the mean tide range is about 80 cm, there has been concern at the increase in the height and frequency of the highest tides ('aqua alta'), which cause flooding in the city area. As will be seen, these are partly the outcome of coastal land subsidence and partly the result of Adriatic storm surges, but an actual rise of sea level is thought to have contributed. In analyses of 7-year running means of the highest annual tides, Pirazzoli (1983) was able to detect maxima in 1931, 1950 and 1969 that coincided with peaks of the 18.6-year 'nodal tide' cycle. Such fluctuations aside, a rise in global mean sea level will carry the highest tides to ascending levels, and it is the maximum, rather than mean, that will command attention.

SEA LEVELS IN THE PAST

The geological record shows that there were major marine transgressions and regressions during Cretaceous and Tertiary times, and it is thought that these were due mainly to tectonic movements of the world's continents. When the land subsided sea level rose to invade the continental margins, and when it was uplifted, sea level fell. These oscillations were followed in the Quaternary by large-scale fluctuations of global sea level accompanying the waxing and waning of glaciers, ice sheets and snowfields which resulted from the successive climatic oscillations that produced the glacial and interglacial phases of the Pleistocene. During glacial phases the Earth's hydrological cycle was interrupted when the climate cooled sufficiently for precipitation to fall as snow, which accumulated as the glacial ice and persistent snowfields that formed in polar and mountain regions. Retention of large amounts of water frozen on land depleted the oceans, and global sea level fell. During the milder interglacial phases the trend was reversed, the water released from melting snow and ice flowing back into the ocean basins to produce a global sea level rise. As will be seen (page 16), these alternations are termed glacio-eustatic. Around 18 000 years ago, when the Earth was in the very cold Last Glacial climatic phase, with a much deeper and more extensive ice and snow cover, global sea level was about 140 m lower than it is now (Ewing 1962). The continental shelves had emerged as wide coastal plains, and the coastline lay well seaward of its present position. The British Isles were uplands on an extensive north-western European land area, and Tasmania stood up from a south-eastern Australian peninsula (see Fig. 81, page 129).

The Earth's climate then began to ameliorate, and as it became warmer the melting of glaciers, ice sheets and snowfields released water into the ocean basins, producing the world-wide sea level rise known as the Late Quaternary marine transgression. Stages in this sea level rise have been determined by dating (mainly by radiocarbon assays) such materials as shells, wood or peat associated with shore deposits found at measured depths in stratified Holocene sediments beneath the sea floor, samples being obtained from cores drilled into these sediments. When the dates are plotted against former sea levels, graphs are produced which trace the sequence of sea level changes relative to the land for particular coastal regions. The resulting curves show that the sea level rise was rapid, averaging just over 1 m per century, between 18 000 and about 6000 years ago. Most studies have deduced an oscillating rise, with pauses and occasional slight regressions, but some have argued for a smooth and steady increase, interpreting the supposed oscillations as statistical aberrations. Discrepancies have been found between Holocene sea level curves from different parts of the world's coastline (Fig. 5), some of which result from the complicating effects of land uplift or depression in coastal regions, and some from actual regional variations in the scale and sequence of sea level changes (see below, page 21); others may be due to errors in the measurement of former shore levels, or in the interpretation and dating of sedimentary deposits.

In Holocene times (here defined as the past 10 000 years) the Earth's climate became more stable and the Late Quaternary marine transgression slackened or came to a halt. Globally, sea level has been relatively stable during the past 6000 years, except on coasts where the land margin has been rising or

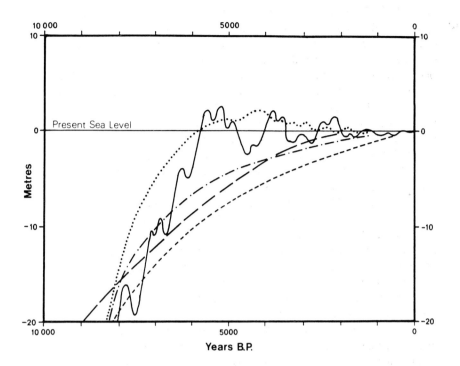

Figure 5 Graphs showing Holocene sea level changes. (———) Global sea level curve according to R.W. Fairbridge; (— —) global sea level curve according to F.P. Shepard; (. . . . . .) sea level curve from New Zealand; (– . – . –) sea level curve from southern England; (- - - - -) sea level curve from New England. For more detail on Holocene sea level curves, see Bloom (1979) and Pirazzoli (1991a)

sinking, but minor oscillations, of the order of 1 m, have been reported from some coasts. The past 6000 years can thus be considered a period of Holocene stillstand, when the relationship between land and sea level has remained stable, but it is acknowledged that some coasts have been emerging or submerging within this period. Pirazzoli (1991a) has assembled the evidence of Holocene sea level changes in a world atlas, which gives information on these geographical variations in the global sea level record.

CAUSES OF SEA LEVEL CHANGES

Several studies have been made of the problems of identifying and explaining sea level changes (e.g. Lisitzin 1974, Tooley 1985, Van de Plassche 1986), and the topic is being investigated by the International Geological Correlation Programme (IGCP) Project 274, led by Dr O. Van de Plassche, which produces annual reports, and will issue a global review in 1993. It will be useful here to summarise the main causes of sea level variations.

A rise of sea level occurs when the volume of water in the ocean basins increases, and a fall when it is reduced: these changes have been termed *eustatic*. Apart from the augmenting of the oceans by the arrival of small quantities of juvenile water supplied from the Earth's interior, primarily from volcanic eruptions, changes in the volume of the oceans are responses to changes in global climate. An increase in atmospheric temperature, for example, results in warming and expansion of the oceans, and sea level rise, whereas if the oceans cool they contract, and sea level falls (Wigley and Raper 1987). The volume of sea water also diminishes as salinity increases, and rises as it freshens. These are known as steric changes.

As has been noted, global climate determines the extent of ice and snow on the Earth's surface. The changes in ocean volume that occurred during Pleistocene glacial and interglacial phases, as a consequence of alternate freezing and melting of the world's glaciers, ice sheets and snowfields, are termed glacio-eustatic. The consequent rises and falls of global sea level culminated in the so-called Postglacial phase, which began about 18 000 years ago, and resulted in the world-wide Late Quaternary marine transgression.

The term 'Postglacial' can be applied locally in deglaciated regions, such as the British Isles, but is unsatisfactory globally in view of the continued existence of polar ice sheets and mountain glaciers, which indicate that the Earth is still in an ice age, in contrast to the ice-free condition that has prevailed through most of geological time, apart from earlier glaciations in the Precambrian and Permian. We are not yet in a Postglacial era, and by the time the remaining land-borne ice sheets, glaciers and snowfields on the Earth have melted they will have released sufficient water to produce a further world-wide marine transgression, raising the level of the oceans by about 60 m (Donn, Farrand and Ewing 1962), and causing widespread submergence and the loss of existing coastal lowlands. The melting of floating ice, as in the Arctic Ocean and the ice shelves of Antarctica, will not increase the volume of the oceans, and will have no effect on sea level.

Sea level can also rise because of the gradual reduction in the capacity of the ocean basins as a result of the accumulation within them of sediment carried from the land to the sea, whether by rivers, melting glaciers, slope runoff, landslides, wind action or coastal erosion. This is a very slow process, termed sedimento-eustatic.

On many coasts, the level of the sea has changed because of tectonic movements, upward or downward, of the coastal land. Around the Mediterranean there is evidence of such uplift in the Oman region of Algeria, the southern Peloponnese and western Crete, and the Izmir region of southern Turkey. Parts of the north coast of New Guinea have been rising intermittently, so that a stairway of successively-formed coral reefs exists on steep coastal slopes. Chappell (1974) deduced that these were the outcome of successive uplifts of coastal land in this area, accompanying a sequence of eustatic sea level oscillations, and endeavoured to separate the two. Similar uplifted terraces have been found in Timor and on the island of Atauro in Indonesia (Chappell and Veeh 1978). Sea level changes on tectonically-active coasts around the Pacific and the Mediterranean have been at least partly due to the rising or sinking of the land margin, either gradually or as a result of sudden earthquakes.

Examples of sea level changes during earthquakes include the Alaskan earthquake of 1964, when parts of Homer Spit, in Kachemak Bay, sank nearly 2 m (Shepard and Wanless 1971), and the Colombian earthquake of 1979, when the coast around Tumaco subsided up to 1.6 m (Herd et al. 1981). In the Rann of Kutch, on the border between Pakistan and India, the 1819 earthquake resulted in an area of about 500 km^2 subsiding beneath the sea. Similar changes have been associated with volcanic activity. The 19th century geologist Charles Lyell observed that the limestone pillars in the Roman market of Serapis, near Naples in Italy, showed a zone about 2 m above present sea level that had been drilled by marine organisms when the sea rose to flood the structure. The land evidently subsided after the pillars were built in Roman times, and was then uplifted; an oscillation thought to be due to evacuation and re-filling of a subterranean lava chamber associated with the nearby volcano of Vesuvius.

Tectonic movements also include the secular isostatic response of the Earth's crust to loading or unloading. In deltaic areas, for example, regional downwarping has occurred as a result of the accumulation of a large sedimentary load, and the coasts of major deltas such as the Mississippi have sectors that show a sea level rise because the level of the subsiding land has not been maintained by sedimentation.

Parts of northern North America and Eurasia that were downwarped beneath spreading ice sheets during glacial phases of the Pleistocene now show an isostatic recovery following their deglaciation and consequent unloading. This adjustment is continuing many centuries after the ice load melted away. Scandinavian coastlines are emerging as a result of this isostatic uplift, which attains a maximum of 1 cm per year on the shores of the Gulf of Bothnia in north-east Sweden and northern Finland (Fig. 6). Successive historical maps show that substantial areas of former sea floor have emerged as new coastal lowlands in the past few centuries. At Skuleberget in north-east Sweden the isostatic uplift has raised a Holocene shore deposit, formed about 8600 years ago, to stand on a hilltop just over 280 m above present sea level.

Another form of isostasy has resulted from the loading and unloading of continental shelves as sea level rises and falls. A rising sea can depress the continental shelf beneath the weight of the deepening water, and the adjacent

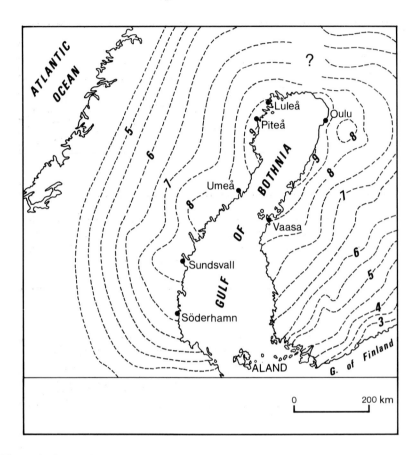

Figure 6   Rates of land uplift around the Gulf of Bothnia, compiled from data from
Swedish surveys of 1886–1905 and 1951–1966 (*Rikets Allmañna Kartveik*, 1971) and
Finnish surveys in 1935 and 1959 (*Atlas of Finland*, 1960). (- - - - -) Contours of land
uplift (mm/year); (●) tide gauges

coast is downwarped as a result. Known as hydro-isostatic subsidence, this
process can augment sea level rise, especially on low-lying coasts that back wide
continental shelves. Similar downwarping can take place where soft sediments,
such as peat, are being compressed beneath the sea floor by the gathering weight
of water as the sea rises (Fig. 7).

Hydro-isostasy blurs the conventional distinction between tectonic movements
of the land and eustatic changes as causes of sea level change. The distinction
disappears when tectonic movements modify the shape and size of the ocean
basins, and thus lead to sea level changes around their coasts. These have

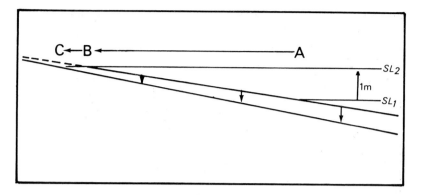

Figure 7   As sea level rises, the weight of the encroaching water may cause hydro-isostatic subsidence, or subsidence due to the compression of soft underlying sediments. This will increase the apparent sea level rise, and result in greater coastline recession (A to C instead of A to B) than on a stable coast

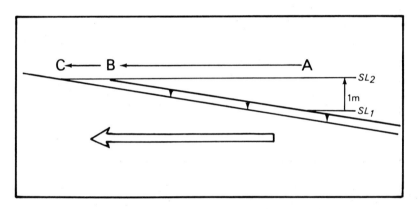

Figure 8   Extraction of underground water, oil or minerals can result in subsidence of coastal land, so that a sea level rise will appear to be larger than on other parts of the coast, with greater coastline recession (A to C instead of A to B)

been styled tectono-eustatic, to distinguish them from the glacio-eustatic and sedimento-eustatic changes mentioned previously.

Sea level can also rise on coasts that are subsiding as a result of human activities such as groundwater extraction, which depletes the aquifers under and around coastal urban and industrial centres (Fig. 8). As underground water is withdrawn, the sediments of the aquifers are consolidated and compressed by the weight of overlying rock formations (and buildings, if any), and the loss in volume results in subsidence of the land surface. In the Venice region, for example, coastal land has subsided by up to 30 cm since 1890, largely because of groundwater extraction. The rate of subsidence has slackened since this groundwater extraction was halted in 1975, but regional tectonic movements and the rising level of the Adriatic Sea still pose problems for Venice. Bangkok (Fig. 9) is one of several other coastal cities where land subsidence has been mapped and correlated with groundwater extraction. The pumping of oil from underground strata south of Los Angeles, California, has led to coastal subsidence, and there have been similar effects near Galveston in Texas and on the Bolivar coast in Venezuela. In the Ravenna region subsidence of up to 1.3 m occurred between 1950 and 1986, due to subsurface compaction following the

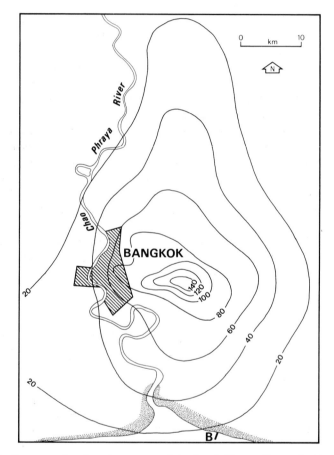

Figure 9 Contours showing the extent of land subsidence (in centimetres) in the Bangkok region between 1933 and 1978, due largely to groundwater extraction and the compression of depleted subsurface aquifers. The subsidence extended to the coastal area (shaded) around the mouth of Chao Phraya River. B, Bangpu Pier. Based on information from the Royal Thai Survey Department

extraction of natural gas as well as groundwater. The drawdown has been most severe over the Ravenna Terra gas field north-east of the city, and there has been increased sea flooding and beach erosion along the Adriatic coastline where it intersects the subsidence bowl. A major impact will occur along this coast if gas removal takes place from pools which underlie the Adriatic coastline (Gambolati et al. 1991). Where oil, natural gas or mineral extraction occur beneath the sea floor any subsidence caused will deepen the ocean basin and result in a lowering of sea level or an offsetting of sea level rise caused by other factors.

Sea level changes have also been influenced by shifts in ocean surface topography that result from geoidal changes, as Mörner (1976) pointed out. The surface of the oceans is undulating, with high and low areas having a total relief globally of the order of 180 m. These bulges and troughs are related partly to gravity patterns and other geophysical phenomena, including tides and the Earth's rotation, and partly to climatic patterns and ocean circulations. In the North Atlantic, tide gauge records show that, when the effects of coastal tectonic movements were excluded, mean sea level rose between 1920 and 1950, the rise being faster on the American coast than in Europe, probably because of variations in sea surface topography related to changes in the Gulf Stream (Pirazzoli 1989). Similar changes have occurred as a result of variations in the distribution of oceanic bulges and troughs, especially during the Quaternary. They are probably responsible for some of the discrepancies between Holocene sea level curves around the world's coastline (Pirazzoli 1991a). Movements of bulges and troughs will be recorded as a sea level rise where high areas move coastward and a sea level fall where low areas move in (Fig. 10). The notion of a global eustatic sea level curve, with equivalent rises and falls around the world's coastline, except for discrepancies resulting from local or regional tectonic movements, had to be abandoned when these geoidal changes were recognised.

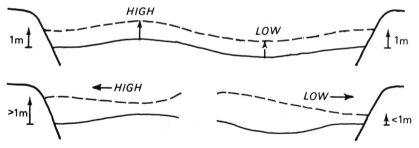

Figure 10 Response to migration of ocean surface topography. The ocean surface shows undulations related largely to gravitational effects and current flow. If a sea level rise is accompanied by landward movement of a high ocean area, it will be augmented on the coast; if sea level rise is accompanied by landward movement of a low ocean area, it will be reduced at the coast

A rise or fall in sea level recorded on tide gauges can occur where artificial structures have been built on the coast. Ports and land reclamation schemes, including the construction of artificial islands, modify coastal and nearshore configuration, and can lead to changes in tide regimes, raising or lowering local sea level, especially in bays and estuaries (Bird and Koike 1986). Changes in nearshore morphology resulting from the dredging and dumping of sediment on the sea floor can have similar effects.

Although there are thus several possible causes of sea level change, the influence of climate is stressed. Climatic fluctuations are thought to be responsible for the oscillations of water level that have taken place in the Great Lakes in North America, during the past century (Fig. 11), and have contributed to the changes in level recorded in the Caspian Sea since 1930 (Fig. 12) (Motamed 1991). In the world's oceans, climatic changes due primarily

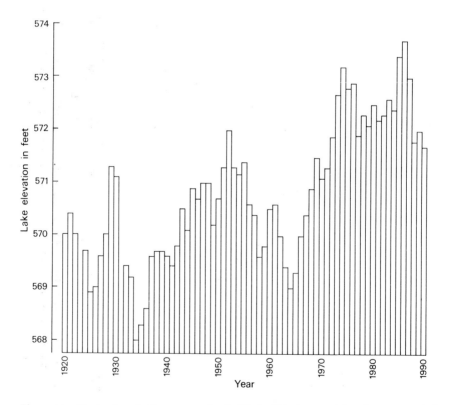

Figure 11   Fluctuations in the summer level of Lake Erie in recent decades, as recorded at Cleveland, Ohio. Data supplied by Dr Charles Carter, University of Akron, and the Great Lakes Environmental Research Laboratory, Ann Arbor, Michigan. Note phases of lake transgression in 1934–1952 and 1963–1974

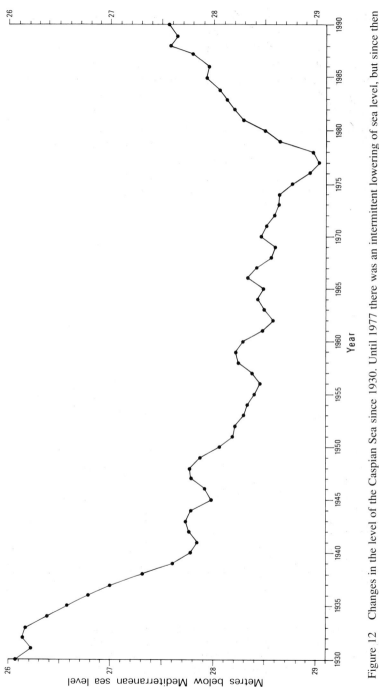

Figure 12   Changes in the level of the Caspian Sea since 1930. Until 1977 there was an intermittent lowering of sea level, but since then there has been a rise of almost 1.5 m. Based on information supplied by P.A. Kaplin, Moscow State University

to variations in solar radiation were responsible for the major fluctuations of sea level during Pleistocene times, and the global marine transgression that culminated in Holocene times. In recent years attention has focused on the probability of human-induced global warming, due to enhancement of the greenhouse effect in the Earth's atmosphere, as a cause of present and future world-wide sea level rise.

THE GREENHOUSE EFFECT AND SEA LEVEL RISE

Measurements initiated during the International Geophysical Year (1957) have shown that atmospheric concentrations of such gases as carbon dioxide, methane and nitrous oxide have been increasing in the Earth's atmosphere (Fig. 13). The carbon dioxide concentration, for example, has risen from 315 parts per million (ppm) in 1958 to more than 350 ppm in 1991, and analyses of past atmospheres trapped in Antarctic ice bubbles have shown that carbon dioxide concentrations were lower, between 280 ppm and 290 ppm, in the 17th and 18th centuries (Pearman 1988).

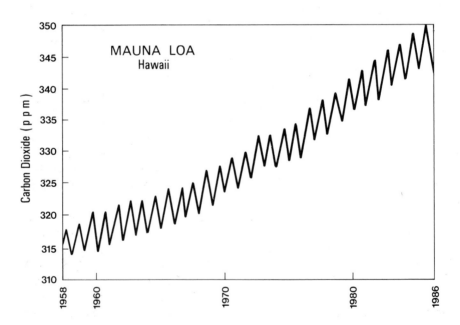

Figure 13   Increasing concentrations of carbon dioxide have been measured in the Earth's atmosphere since monitoring began in 1958. Similar graphs have been obtained for concentrations of other greenhouse gases, notably nitrous oxide, methane and chlorofluorocarbons

The increasing proportion of these greenhouse gases is attributed to the growth of atmospheric pollution resulting from industrial and agricultural activities, including emissions from motor vehicles, which have generated them in increasing quantities, especially since the Industrial Revolution (Jones and Henderson-Sellers 1990). The burning of fossil fuels (coal, oil and natural gas) is returning to the atmosphere carbon dioxide that was withdrawn from earlier atmospheres by plant photosynthesis and retained in swamp forests that became fossil fuel deposits in the geological past.

Existing global temperatures are maintained by the greenhouse effect, whereby the atmosphere intercepts some of the solar radiation reflected back into space from the Earth's surface, and so retains some heat: the Earth would otherwise be much colder. The increasing proportion of greenhouse gases will enhance this effect, making the Earth's atmosphere more opaque to reflected radiation, so that a higher proportion of the solar energy received at the Earth's surface is retained. The outcome will be global warming.

It has been predicted that atmospheric concentrations of the greenhouse gases will double during the coming century, resulting in an increase of 1.5°C to 4.5°C in the mean temperature of the lower atmosphere. Such warming will cause an expansion of the volume of near-surface ocean water (the steric effect), and partial melting of the world's snowfields, ice sheets and glaciers, releasing water into the oceans. The slight decrease in sea salinity will also augment ocean volumes. The outcome will be a world-wide sea level rise (Barth and Titus 1984).

There have been various estimates of the scale of this new marine transgression. Analyses carried out by the Environmental Protection Agency (EPA) in Washington led Hoffman (1984) to predict that a global sea level rise would begin slowly and accelerate to attain 0.24–1.17 m by the year 2050, and 0.56–3.45 m by the year 2100. According to these predictions the mean level of the oceans will rise 1 m by the year 2045 (high scenario), or by the year 2140 (conservative scenario); the most likely (intermediate) scenario was thought to be an accelerating rise attaining 1 m over the coming century.

This is consistent with the 1987 prediction by the Climatic Research Unit, University of East Anglia, that the oceans will rise 12–18 cm above their present level by the year 2030 (Raper, Warwick and Wigley 1988), but more conservative than the prediction from the Villach Conference, held in Austria in October 1985, that global sea level would rise 20–140 cm by the year 2030, due largely to thermal expansion of the oceans. Somewhat higher scenarios have been presented by those who expect a rapid melting of a substantial part of the West Antarctic ice sheet, which is currently grounded on the sea floor, but this is not likely to occur within the next few centuries. Budd (1988, 1991) calculated that melting of the part of the West Antarctic ice sheet that stands above present sea level, together with more rapid downflow of ice from the

hinterland, could raise global sea level by up to 3.5 m over the next thousand years.

The most recent predictions of sea level rise due to the greenhouse effect, based on analyses presented to the Intergovernmental Panel on Climatic Change (IPCC) in 1990, are that global mean sea level will rise 20 cm by the year 2030, and 65 ($\pm$35) cm by the end of the next century. This was the most likely scenario, accompanied by high and low estimates as shown in Fig. 1, page 3 (Houghton, Jenkins and Ephraums 1990). Even if concerted international action halted the discharge of greenhouse gases into the atmosphere by the year 2030, global sea level would still continue to rise, attaining about 40 cm by the year 2100, and proceeding for many more decades before it stabilised. Theoretically, global atmospheric management could restore the conditions which existed during the 20th century, and return the sea to its present level. Meanwhile, it is useful to consider the possible effects of a global sea level rise of 1 m over the next 100–150 years, and to refine these as more accurate predictions come to hand.

Depletion of the world's upper atmospheric ozone layer, which intercepts much of the ultraviolet radiation arriving from the sun, is thought to be due to the effects of chlorine monoxide produced from chlorofluorocarbon (CFC) emissions, generated by aerosol propellants, refrigerators and various industrial processes. Ozone depletion leads to an increase in ultraviolet light penetrating to the Earth's surface, where it is harmful to the growth and health of plants and animals, and damaging to human eyes, skin and immune systems. In addition to depleting ozone, CFCs also contribute to the enhanced greenhouse effect. The Montreal Protocol of 1987 paved the way for international agreements to achieve the reduction and eventual cessation of CFC emissions, but this will have only a minor effect on the predicted global warming and sea level rise.

Forecasting of actual sea level changes at various places on the world's coastline is complicated by the various factors previously discussed, especially continuing tectonic movements in coastal regions. Where the land is subsiding the rate of sea level rise in the coming century will be faster than the global average. The same will be true where the additional water load hydro-isostatically depresses the submerging land margin and thus augments sea level rise. The spatial variability of ocean surface topography and the expected migrations of bulges and troughs make it unlikely that the rise of sea level will be equivalent around the world's coastline, even on sectors where the land margin remains stable (Mörner 1985, Pirazzoli 1986a). Land uplift resulting from isostatic recovery following deglaciation of the Scandinavian region will continue to raise the coasts of Finland and northern Estonia, as southern Estonia and the south coasts of the Baltic Sea subside. A global sea level rise of 1 m over the next 100–150 years will be diminished by continuing isostatic uplift in northern Sweden and Finland: in the Oulu region, where the land is rising about

1 cm per year, there will be no apparent change of sea level after this period. In southern Finland and northern Estonia isostatic uplift will reduce the global sea level rise by several centimetres, whereas in southern Baltic regions subsidence will augment the relative rise of sea level.

The scale of sea level rise will also vary where there are changes in tidal amplitude caused by accompanying modifications of coastal and nearshore configuration. Where the tide range now diminishes into embayments with narrow entrances (as in Port Phillip Bay, Australia) a sea level rise is likely to increase it, and where it is presently amplified towards the heads of funnel-shaped inlets (as in the Bay of Fundy, Canada and the Gulf St Vincent and Spencer Gulf in South Australia) a sea level rise could be reduced below the global average (Fig. 14). The modifications will generally be small, except where sea level rise brings the tidal system closer to resonance (i.e. when tide wave length becomes similar to that of the basin into which it flows, determined by coastal and sea floor configuration), in which case they could be magnified.

In The Netherlands it has been estimated that a 1-m global sea level rise will increase the height of high tides in the Wadden Sea by up to 1.06 m. Misdorp et

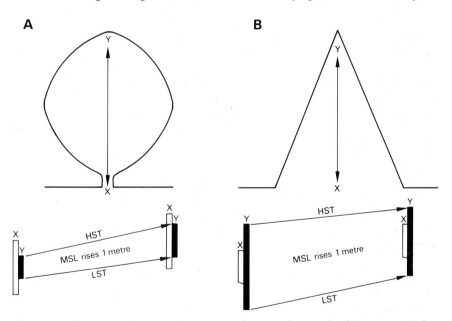

Figure 14   Sea level rise may be accompanied by changes in tide range: (a) in embayments with narrow entrances, where the tide range at the head of the bay (Y) is likely to increase compared with that at the entrance (X) because of increasing tidal ventilation; (b) in funnel-shaped gulfs, where tide range at the head of the gulf is likely to be reduced (Y) because of diminishing tidal ventilation. HST, high spring-tide; MSL, mean sea level; LST, low spring-tide

al. (1989) found that tide ranges had already increased there by up to 15 cm in the past century, high tides having risen while low tides remained unchanged, so that mean sea level had ascended. These changes were thought to be the outcome of an enlargement by current scour of the cross-sectional areas of tidal inlets between the barrier islands that separate this shallow area from the North Sea.

IS SEA LEVEL RISING?

Uncertainty persists over whether the greenhouse effect has already produced global warming and a sea level rise. The various greenhouse gases produced by agricultural and industrial activities have been accumulating in the atmosphere for several centuries. As has been noted, comparisons of existing concentrations of carbon dioxide with those preserved in Antarctic ice bubbles dating from the pre-industrial era have shown an increase of nearly 30% during the past 200 years, and this should have had a measurable effect on atmospheric temperatures.

Some climatologists claim to have detected a rise of about $0.4°C$ in global atmospheric temperatures since the beginning of the present century (Hansen et al. 1984), but others have been more cautious, noting for example that there have been 'heat island' effects on temperatures recorded in urban and industrial areas where many meteorological stations are situated. Even a $0.4°C$ rise is less than expected, according to climatological models which show that the observed increases in greenhouse gases should have raised temperatures between $1°C$ and $2°C$ over this period. Factors that may have complicated the situation include the 'feedback' effects of increasing cloudiness and slow deep-ocean circulations. Global temperatures have also responded to variations in solar radiation, notably the 8–14 year sunspot cycle, and the earlier reductions that caused the Pleistocene ice age. One possibility is that the Earth's atmosphere is actually in a cooling phase, which has been obscured by the enhancement of the greenhouse effect, so that the temperature rise has been held back. Although most climatologists expect global warming as a consequence of the enhanced greenhouse effect, a few have predicted that increasing cloudiness will impede temperature rise at the Earth's surface, producing a damp and sombre global climate, without significant warming. For the purposes of this study the majority view is accepted.

If warming of the atmosphere has already occurred, global sea level should have been rising around the world's coastline. Evidence for such a rise may be sought from long-term tide gauge records. Among the longest continuous records available are those from San Francisco (USA), Brest (France) and Sydney (Australia), each of which has shown fluctuations with a generally upward trend. Several attempts have been made to determine global trends in sea level from statistical analyses of long-term tide gauge records (Barnett 1984, Gutenberg

1941, Valentin 1952), and there is wide support for the idea that global sea level, having risen 10–15 cm during the past century, is now rising by about 1.2 mm per year (Fairbridge 1966, Fairbridge and Krebs 1962, Gornitz and Lebedeff 1987).

However, as Pirazzoli (1986a) demonstrated, tide gauge records from around the world over recent decades have shown variations in mean sea level trends. Of 229 tide gauge stations with at least 30 years' reliable recent records 63 (28.5%) showed a mean sea level rise >2 mm per year, 52 (22.5%) 1–2 mm per year, and 47 (20.6%) < 1 mm per year, the remaining 65 (28.5%) having shown a mean sea level fall. As over 70% of the records showed a positive trend a global sea level rise seems likely, but such a conclusion may be premature in view of the uneven distribution of the 229 stations, only 6 of which are in the southern hemisphere (Fig. 15). Emery and Aubrey (1991) extended this review, using 664 tide gauge stations, 65% of which had at least 30 years' records. Their analysis of the key 98 stations showed that sea level fall was confined to sectors known to be rising tectonically or isostatically. As Fairbridge (1987) pointed out, reliable global averages cannot be obtained when the tide gauge data show such strong northern hemisphere mid-latitude clustering. It is expected that the Global Sea Level Observing System (GLOSS), which is developing a much more representative network of tidal stations around the world's coastline (Fig. 16), will provide more accurate information in the next few decades (Pugh 1987a, Woodworth 1991), both globally and regionally (Pugh 1991), but by then the sea level changes due to the enhanced greenhouse effect may be obvious.

Meanwhile, the suggestion that mean sea level is rising globally should be treated with caution, bearing in mind the geographical variations shown by Holocene fluctuations of sea level (Fig. 17). Sea level changes over the past few decades should be considered against the background of factors listed previously, which are known to have influenced Holocene sea level trends.

The most obvious of these are tectonic movements. Sea level has been falling on coasts where isostatic land uplift has followed Late Quaternary deglaciation, as in parts of Scandinavia, northern Canada and Alaska, and where the land is rising tectonically, as in northern New Guinea, some Indonesian and Philippine islands, and parts of the Japanese coast (Pirazzoli 1987). Tectonic subsidence of the ocean floor, by enlarging the capacity of the ocean basins, would also lead to a fall in sea level, or at least a partial offsetting of sea level rise due to other factors, but our knowledge of vertical movements of the sea floor (as distinct from horizontal movements known to be in progress at the margins of tectonic plates) is poor, and will remain so until satellite monitoring of the Earth's crust is established in relation to a centre-of-the-Earth datum.

Sea level has been rising relative to coasts where the land margin has been subsiding, as on the Gulf and Atlantic seaboards of the United States, the south and south-east coasts of Britain, The Netherlands and north Germany, north-

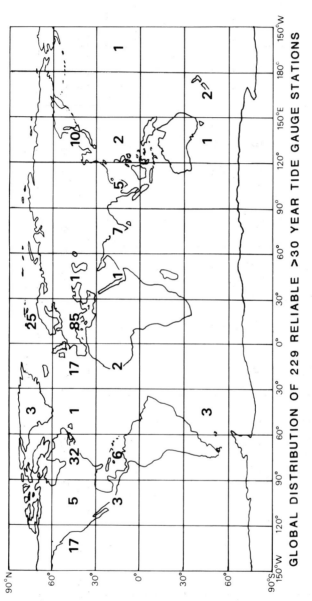

GLOBAL DISTRIBUTION OF 229 RELIABLE >30 YEAR TIDE GAUGE STATIONS

Figure 15 Global distribution of sea level trends based on records from tide gauge stations that have operated continuously for more than 30 years, from data summarised by Pirazzoli (1986a). He noted that of the 229 stations, 65 (28.5%) had shown a mean sea level rise of more than 2.0 mm/year; 52 (22.5%) 1.0–2.0 mm/year; 47 (20.5%) less than 1.0 mm/year, while 65 (28.5%) had recorded a sea level fall. Most of the tide gauge stations are in western Europe and North America, and only 6 of them are in the Southern Hemisphere. More reliable analyses of trends of sea level change will be possible when the GLOSS network (Fig. 16) has been in operation for at least two decades

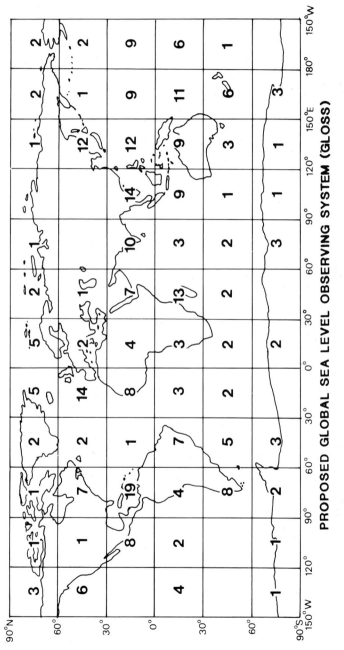

PROPOSED GLOBAL SEA LEVEL OBSERVING SYSTEM (GLOSS)

Figure 16 The proposed network of tide gauge stations being established by the Global Sea Level Observing System (GLOSS; IOC, Paris, October 1986) to improve monitoring of sea level changes

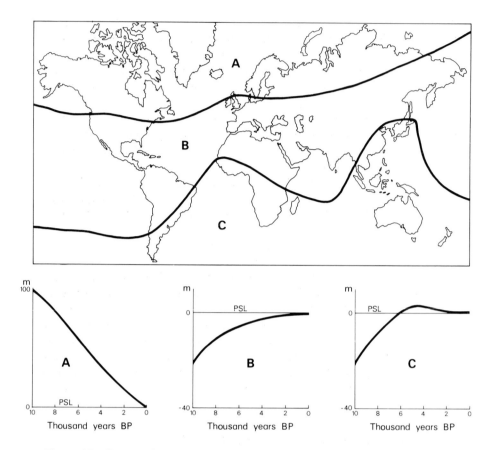

Figure 17   Graphs of sea level change during the Holocene fall into three main categories: (a) in high northern latitudes, where land uplift has led to a falling sea level; (b) in middle latitudes, where the sea has risen at a slackening rate to attain (but not exceed) present level; (c) in much of the southern hemisphere and south-east Asia, where the sea rose above its present level between 3000 and 6000 years ago, and has since fallen back. In detail there are many variations within the three categories, related especially to local uplift or depression of coastal land, but also to geodetic changes in ocean level. PSL, present sea level. Based on a diagram by P.A. Pirazzoli

eastern Italy, and several other areas. There has been a growing awareness of the extent of subsiding coasts: Fig. 2 (page 5) shows the location of sectors of the world's coastline that have been subsiding in recent decades, as indicated by evidence of tectonic movements, increasing marine flooding, geodetic surveys, and groups of tide gauges recording a rise of mean sea level greater than 2 mm per year over the past 30 years. It is probable that further research will increase the number and extent of these subsiding sectors. Some are in areas of isostatic downwarping around major deltas; others are at least partly the outcome of human activities, notably groundwater extraction, as in the Venice and Bangkok regions, and oil extraction, as in southern California. Some of the sectors are too small to be accurately portrayed on a world map: the actual areas of subsidence around Bangkok, Thailand are shown in Fig. 9 (page 20) and those in the vicinity of Niigata, Japan in Fig. 70 (page 116).

Evidence of sea level rise from tide gauge records has been further complicated by fluctuations due to short-term climatic phenomena such as the ENSO and the other oceanic fluctuations. These have to be extracted from tide gauge records before any secular sea level rise can be demonstrated. In the United Kingdom, Woodworth (1987) found from tide gauge records that a general rise in sea level, relative to western Europe, had been interrupted by a sharp fall in the mid-1970s, probably due to modifications in the North Atlantic Oscillation.

The balance of the evidence suggests that a global sea level rise is probably in progress, but more extensive monitoring is required to confirm this. There is still a possibility that the so-called contemporary world-wide marine transgression of about 1.2 mm per year has been overestimated, or that more complete global studies will show that there have been geographical variations in the nature and scale of sea level change during the past century similar to those indicated by Holocene sea level curve discrepancies. It is, nevertheless, expected that an accelerating global sea level rise due to the enhanced greenhouse effect, accompanied by fluctuations caused by the other factors that have been mentioned, will become obvious on the world's tide gauges during the next few decades.

BIOLOGICAL ZONATIONS AS SEA LEVEL INDICATORS

The limitations of tide gauge records and geodetic surveys as indicators of sea level changes have prompted consideration of biological zonations as sea level indicators (Kidson and Heyworth 1979). Evidence of sea level changes may be obtained from repeated surveys of the levels of marine organisms such as oysters, barnacles, mussels, algae and kelp, where these are found in vertical zonations, occupying specific levels, encrusting cliffs, rocky shores, sea walls and pier supports. Their levels are correlated with the depth and duration of marine submergence with the sea at its present level (Lewis 1964), and can be expected to move up or down in relation to a rise or fall of sea level (Fig. 18). They should move upward on cliff faces, stacks and rocky protrusions, as well as on artificial structures, as sea level rises, and landward across shore platforms and intertidal outcrops (Fig. 19).

Evidence that they can do so has been reported from the subsiding coast of south-east Florida by Wanless (1982), who showed that horizons of oysters and

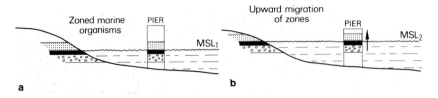

Figure 18   Where zonations of marine organisms have developed on rocky shores or artificial structures, with specific relationships to present sea level, they will migrate upwards in response to a sea level rise. It is possible that some zoned organisms can be used to supplement evidence from tide gauge records as indicators of changing sea level. (a) Present situation. (b) Response to sea level rise

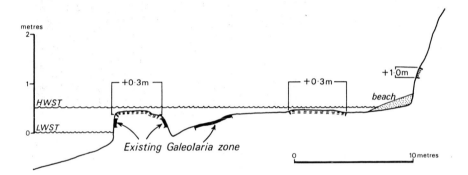

Figure 19   The calcareous tubeworm *Galeolaria caespitosa* occupies part of the intertidal zone between low water spring-tide (LWST) and high water spring-tide (HWST) on rocky shores in south-eastern Australia. The diagram shows how it is expected to migrate upwards on to shore platforms with a sea level rise of 30 cm, and on to the cliff face with a sea level rise of 1 m

barnacles moved 15 cm upwards on concrete piling at Miami Beach between 1949 and 1981, consistent with a rise of mean sea level registered on nearby tide gauges. It may be possible to detect such migrations with reference to historical photographs where mean sea levels can be determined in relation to fixed features such as steps or decking, and it would be useful to document existing levels of zoned organisms as a basis for future measurements. The most suitable organisms are readily identifiable plants or animals that occupy particular parts of the intertidal zone with consistent, well-defined upper and/or lower boundaries that can be correlated with particular stages of the tide, preferably at or close to mid-tide level. Indicator organisms should be able to migrate upwards as sea level rises by rapidly colonising higher levels that are either untenanted, or occupied by other plants or animals that can be displaced or overrun.

   An example from south-eastern Australia is the calcareous tubeworm *Galeolaria caespitosa*, which occupies a well-defined intertidal horizon, and is

Figure 20   Cushions of *Galeolaria caespitosa* on pier supports at Rosebud, on the shores of Port Phillip Bay. The upper limit of these cushions is likely to be raised in response to a rising sea level

best developed on sites sheltered from strong wave action and abrasive sand movements. It forms either a thin layer, or cauliflower-like encrustations (Fig. 20) similar to the 'trottoirs', accretionary ledges of coralline algae that protrude from some Mediterranean rocky shores. The vertical range of *Galeolaria* is typically 20–40 cm, with an upper limit close to mid-tide level, irregular on sites exposed to strong waves and variable swash and spray, but horizontal on sheltered sites such as the inner sides of harbour walls. Within Port Phillip Bay, where tide gauges show little evidence of any recent sea level change, the upper limit of *Galeolaria* has shown no consistent indication of an upward movement in recent years (Bird 1988a).

Coral growth on intertidal reef flats is likely to be stimulated by a rising sea level, but, as will be seen, there has so far been little or no evidence of a general revival of coral growth on these surfaces. The evidence is discussed in Chapter 3.

OTHER EFFECTS OF GLOBAL WARMING

As a global sea level rise proceeds, it will be accompanied by other changes resulting from warming produced by the enhanced greenhouse effect. These are likely to modify the various oscillations, both cyclic and irregular, that already occur in the world's oceans. An indication of possible climate changes is given by the El Niño responses: droughts in Australia and Africa, heavy rain and river flooding in China, Peru and Ecuador. Geomorphological and ecological responses to rising sea level will be complicated by the effects of a changing climate. As warming proceeds some regions are expected to receive more rainfall while others become drier (Henderson-Sellers and Blong 1989, Houghton, Jenkins and Ephraum 1990, Tegart, Sheldon and Griffiths 1990). Some rivers will carry more sediment because of heavier catchment runoff, but many already have increased sediment loads because of the depletion of catchment vegetation by increasing aridity, or by clearance, especially where forest areas have been adapted for agricultural use, and soil erosion has ensued, or where mining has led to increasing sediment flow downstream. The capacity of rivers to move sediment down to the coast will increase as fluvial discharge rises, but the increase may be curbed by natural or human-induced revegetation of their catchments. However, there are some areas, notably around the Mediterranean, where fluvial sediment yields are now diminishing because parts of the upper catchments have been almost completely stripped of soil and weathered mantles, and are now bare and rocky.

Where coastal waters become stormier as sea level rises, increased amounts of sediment will be derived from the more rapidly eroding cliffs and foreshores. This could result in nearby or alongshore deposition that at least partly offsets the land losses that would otherwise occur as submergence proceeds.

As well as raising temperatures in coastal regions, global warming will cause a migration of climatic zones, tropical sectors expanding poleward as the domain of arctic coasts shrinks. Tropical cyclones may become more frequent and severe. They will extend into higher latitudes, bringing storm surges and torrential downpours to coasts that now lie outside their range. These effects could be more damaging than the direct impact of sea level rise (Daniels 1992). Coasts already subject to recurrent storm surges will have still more frequent and extensive marine flooding as submergence proceeds, and more severe erosion and structural damage where larger waves reach the coast through deepening nearshore waters.

Some coastal regions will become wetter, others drier. Where rainfall increases there will be more frequent and persistent river flooding, and the water table rise will be added to that caused by the rising sea level. Some low-lying parts of coastal plains will become permanent swamps or lakes, the salinity of which will depend on interactions between increasing marine incursion (and perhaps an upward movement of subterranean salt) and any offsetting effects of augmented rainfall and freshwater runoff.

With increased rainfall, coastal vegetation may become more luxuriant; with drier conditions it is likely to be depleted. Coastal regions that become drier will also have more extensive marine salinity penetration into both surface and underground water. Coastal lagoons will become more brackish, and some may dry out as saline flats. Desiccation will also reduce vegetation cover, and allow wind action to be more erosive, especially in coastal dune areas, whereas more vigorous and extensive vegetation will tend to stabilise dunes that are now bare and drifting. Vegetation and fauna may also be impoverished by the effects of increasing ultraviolet radiation resulting from ozone depletion.

Most marine plants and animals have specific latitudinal ranges, usually depending on maximum or minimum sea temperatures, which will rise as global warming proceeds. Thus mangroves, now largely confined to tropical coasts, will extend their range poleward along coasts when there are suitable habitats, and corals will extend northward and southward beyond their present latitudinal limits, providing they have the capacity to establish on suitable substrates. Many temperate salt marsh species will also migrate poleward, whereas the distribution of kelp, a plant restricted to cooler waters, will contract to higher latitudes.

Global warming is also likely to affect coastal ecosystems by way of thermal stress, a factor already noted where coral bleaching has occurred (see page 93). Vegetation growth could be impeded by higher temperatures, and coastal ecosystems will be adversely affected by increasing ultraviolet radiation. On the other hand, increasing atmospheric carbon dioxide should enhance photosynthesis and plant growth. The maintenance of coastal biodiversity will require the management of coastal reserves to ensure that threatened species are not extinguished by global warming and sea level rise.

CONCLUSIONS

There is no doubt that the proportions of greenhouse gases in the Earth's atmosphere have been increasing, and it is very likely they will lead to global warming and a world-wide sea level rise. However, many other factors influence climatic variations and sea level changes. It is not yet certain that a global sea level rise is taking place; if it is, it may not be due entirely to an enhanced greenhouse effect. While an accelerating sea level rise is expected during the next few decades, it is not yet possible to be sure of its dimensions, or to assess how far and how quickly it could be curbed by restraining atmospheric pollution.

Many coasts have already been much modified by human activities, both directly in the form of artificial structures and land reclamation schemes and indirectly through impacts on the hydrology and ecology of coastal environments. Construction of dams on rivers has reduced water and sediment yield to coastal areas, and is already responsible for coastline erosion, especially on deltas. Many coastal wetlands have been drained, and coastal lagoons modified. Coastal dunes previously held by vegetation have been liberated when the plant area has been reduced by cutting, burning and grazing. Ecological changes have resulted from human adaptations of the coastal environments, including pollution of coastal waters, and some areas of coastal subsidence are partly the outcome of the extraction of groundwater, oil or natural gas. Changes in progress on many coasts are at least partly the outcome of such human activities, and additional changes will result from a sea level rise due to the enhanced greenhouse effect, which is also a human impact.

Many coastal cities are threatened by the prospect of a global sea level rise. Venice, Shanghai, Bangkok, Hong Kong and Tokyo have been frequently cited as the most vulnerable to the effects of submergence and storm surges.

Chapters 3 and 4 will examine the geomorphological and ecological changes likely to occur if a sea level rise of the order of 1 m takes place, and the possible human responses to such changes.

Chapter Three

Sea level rise and coastline changes

GENERAL EFFECTS OF A RISING SEA LEVEL

Most of the natural features seen on the world's coastline have formed during the period of relatively stable sea level (the Holocene stillstand) that came as a sequel to the world-wide Late Quaternary marine transgression. Exceptions are found on coasts that retain a landform legacy from the Pleistocene, such as the slope-over-wall cliffs of the Atlantic coasts of north-west Europe, which retain slopes mantled by solifluction deposits of Pleistocene periglacial origin, or the high glaciated cliffs that border fiords, which are ice-scoured slopes rather than features of marine origin. On coasts that have been rising as a result of tectonic movements during the past 6000 years, erosion has been impeded and deposition aided by a falling sea level, while on coasts that have been subsiding, erosion has been aided and deposition impeded by a rising sea level. Where the Holocene sea level stillstand has endured for up to 6000 years there has been time for coasts to develop such features as broad shore platforms fronting cliffs, wide coastal plains with multiple beach or dune ridges, coastal barriers and barrier islands fronting lagoons, prograded deltas, and extensive intertidal zones partly occupied by salt marshes or mangroves.

Coastal landforms continue to change as erosion and deposition proceed in response to such processes as wave and wind action, tidal and other currents, weathering, runoff, and gravitational mass movements. Changes particularly noticed are those which occur during extreme events, such as earthquakes and storm surges, when dramatic erosion takes place on some coasts and rapid deposition on others. Monitoring of coastline changes is needed to determine the nature and rates at which such changes are taking place, to provide a basis for assessing what will happen as sea level rises (Bird 1985).

It is necessary to consider first the changes that will take place in the absence of any human response, and then to decide how these may be modified by human activities. If global sea level rises in the manner predicted (intermediate scenario, Fig. 1, page 3) there will obviously be extensive marine submergence of low-lying coastal areas. High and low tide lines will advance landward, and at least part of the present intertidal zone will become permanently submerged. On steep hard rock coasts where there is little or no marine erosion in progress, and on solid artificial structures such as sea walls, a sea level rise of 1 m will simply raise high and low tide lines to this contour above their present levels.

It is possible that there will be a slight increase in tide ranges around the world's coastline as the oceans deepen, the rise that actually occurs being modified as tidal amplitude is adapted to the changing coastal and nearshore configuration. On many coasts the extent to which the high tide line moves landward will be augmented by an increase in erosion as nearshore waters deepen, and larger and more destructive waves break upon the shore. As sea level rises erosion will begin on coasts that are at present stable, and accelerate on coasts that were already receding. On coasts that had been prograding the seaward advance will be curbed, and erosion may begin (Carter and Devoy 1987): on the coasts of the Caspian Sea the proportion eroding was 10% in 1977, but has increased to 39% during the ensuing sea level rise (Fig. 12, page 23).

Erosion will increase still further where the climatic changes that accompany the rising sea level lead to more frequent and severe storms, generating surges that penetrate farther than they do now. The extent of erosion will also depend on how the nearshore sea floor is modified by the rising sea as the coastline moves landward. Although there will generally be deepening of nearshore waters and a consequent increase in wave energy approaching the coastline, some sectors may receive an augmented sediment supply, derived from increasing fluvial or alongshore sources, and this could maintain, or even shallow, the nearshore profile as the sea rises. On such coasts wave energy will not intensify, and there may be little, if any, coastline erosion; there may even be some progradation.

The various sea level scenarios propose an accelerating rise, with no prospect of stabilisation. If the sea continues to rise, coastal erosion is likely to accelerate and become more widespread as any compensating sedimentation declines. It will encounter, re-shape, and in due course submerge the 'raised beaches' and other shore features that formed at higher sea levels during Pleistocene interglacial phases. Revival at higher levels of features similar to those now seen on the world's coastline, with some sectors stable and others prograding, can occur only after a new sea level stillstand has become established.

Stabilisation of global sea level could be achieved by human intervention, a decision to manage the atmosphere and the oceans to restore a balance between global temperatures and ice volumes, but as long as global warming continues there is no prospect of such a stillstand. Global sea level could rise about 60 m, when all the world's glaciers, ice sheets and snowfields have melted and the water contained in them has flowed back into the oceans (Donn, Farrand and Ewing 1962). Even then, if warming continues, sea level will go on rising because of thermal expansion, until the countering effects of soaring evaporation finally overcome the rising trend. The eventual prospect of boiling oceans, an atmosphere on fire, and yet another lifeless planet is well beyond the scope of the present enquiry.

Because of the complexity of coastal environments, and their alongshore variations, a great deal of sector-specific research will be needed to predict the extent of the geomorphological and ecological changes that will occur as sea level rises (Mehta and Cushman 1989). Coastal landforms and ecosystems have been subject to many changes, both natural and due to human activities, in recent centuries: these changes are still in progress, and will continue even if the sea were to remain at its present level. It becomes a question of predicting the coastal changes that will result from a sea level rise, and distinguishing them from those that would have occurred had the global environment remained stable (Bird 1988b).

Attempts to predict the extent of coastal changes as sea level rises have run in to a number of difficulties. Knowledge of the interactions between coastal hydrodynamics and sediment flow patterns has so far proved inadequate for generating reliable predictive models. It is necessary to take account of accompanying changes in climate, which may increase or reduce the incidence and effects of runoff and river flooding, storminess, wind action, weathering, and biological processes. Some attempts have been made to forecast the nature and extent of coastline changes that will result from a sea level rise on particular coasts, notably in the United States (Titus 1986b), The Netherlands (De Ronde 1991), the British Isles (Boorman, Goss-Custard and McGrorty 1989) and New Zealand (Gibb and Aburn 1986). Evidence of geomorphological and ecological changes that will take place as sea level rises can also be sought from those parts of the world coastline that have been subsiding, and which therefore already show responses to a relative rise of sea level.

In assessing the nature and extent of possible changes, it is useful to classify coasts into several categories: steep and cliffed coasts, beach-fringed coasts, deltaic coasts, swampy coasts, estuaries and coastal lagoons, intertidal and nearshore areas, coral reefs and reef islands, and artificial (man-made) coasts. These will be considered in subsequent sections, dealing first with relevant aspects of the evolution of existing features, then with the possible (theoretical) changes in response to a sea level rise, with evidence of what has happened to these features where subsidence has taken place and, where applicable, evidence from the Holocene stratigraphic record of the changes that occurred during the Late Quaternary marine transgression.

Regional and national predictions of the effects of a rising sea level require that an assessment be made of the proportions of each of these coastal landform categories. A Geographical Information Systems approach can be used to approximate these from existing maps and air photographs, but there are dangers in generalisation. Beaches, for example, vary in the nature and calibre of their component sediments, their hinterland and nearshore features, their configuration and aspect in relation to waves and winds, and tidal regimes, all of which must be taken into account in classifying them on the basis of their response to sea level rise. The extent of cliffed coasts can be mapped readily, but the ways in which they have been changing over the past few centuries are difficult to establish, and their response to sea level rise will vary with their lithology and structure, with hydrological conditions, weathering and induration effects, and the features of the adjacent sea floor. For many countries, knowledge of existing coastal geomorphology and ecology is still at a reconnaissance stage (Bird and Schwartz 1985), and more detailed surveys are now required to provide the data necessary for making predictions of sea level rise response. In due course, Geographical Information Systems will be used to integrate sea level rise scenarios with topographic and socio-economic variables, as outlined in general by Shennan and Tooley (1987) and with reference to the Tees estuary by Shennan and Sproxton (1991). Examples of modelling that could be developed for sea level rise scenarios are given by Krysanova et al. (1989), Wroblewski and Hoffmann (1989), and Constanza, Sklar and White (1990). As Mimura, Isobe and Nadaoka (1991) pointed out, the need for appropriate response strategies to sea level rise will require more qualitative impact assessments than are currently available. Meanwhile, Sestini (1992) has tabulated systems interactions on submerging coasts, identifying the various components (Fig. 21).

STEEP COASTS: CLIFFS AND BLUFFS

Cliffs have formed on coasts where the Late Quaternary marine transgression brought the sea alongside high ground, with wave action sufficiently strong to be erosive, cutting steep to vertical features that expose rock formations. Cliff morphology is related to geological structure and rock type, and cliffs are generally bolder on coasts exposed to strong wave attack, especially where marine erosion is concentrated within a thin vertical zone by a small tide range. Receding cliffs are generally fronted by shore platforms, which typically decline seaward from high tide level at the cliff base out to below low tide level. Shore platforms are generally wider where the tide range is large, or where the coast is exposed to strong wave attack. On some coasts they are almost horizontal, standing at various intertidal levels, and determined either by structure (e.g. horizontal outcrops of resistant rock) or weathering (e.g. disintegration or solution down to a specific level, such as mean low tide on limestone outcrops). Elsewhere there is either a rugged rocky shore topography, or a plunging cliff descending into deep water.

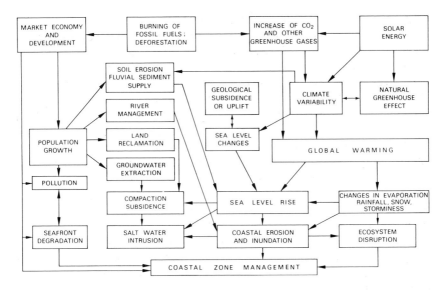

Figure 21 Systems interactions related to submerging coasts. Modified from Sestini (1992)

The present features of cliffs and shore platforms are largely a consequence of the long phase of relatively stable sea level around much of the world's coastline during the past 6000 years. There has been little cliffing on emerging coasts, such as those of the northern Baltic region, where wave erosion has been withdrawing seaward across the emerging sea floor. Cliffing is also limited between the terraces that mark the intermittent uplift of the north coast of New Guinea, which are evidently emerged coral reef fronts (Chappell 1974). The features of cliffs and shore platforms on submerging coasts will be discussed subsequently.

Existing rates of cliff retreat vary with such factors as rock resistance, structure, the presence or absence of shore platforms or nearshore reefs, exposure to wave action, and tide range (Sunamura 1992). Where soft materials, such as volcanic ash and glacial drift, are exposed to strong wave attack, cliffs have been retreating by up to hundreds of metres in a year; at the other extreme cliffs in hard rocks such as massive granite have shown little change over many centuries. Cliffs cut in seaward-dipping formations are apt to retreat more rapidly than those on horizontal or landward-dipping strata because of the greater incidence of rock falls and slumping down seaward-inclined bedding planes. A shore platform or nearshore reef tends to retard cliff recession by diminishing wave energy, in comparison with adjacent sectors on similar formations where such protection is lacking. In the same way, cliffs on a sector of coast exposed to strong wave action retreat more quickly than those facing in other directions.

On a particular rock formation, a large tide range, by dispersing wave energy over a broad intertidal zone and bringing waves to the cliff base only briefly at high tide, results in more gradual recession than where a small tide range concentrates wave energy and maintains a more consistent attack on the cliff base.

Steep coasts may consist of bluffs mantled with weathered material and soils, held in place by a vegetation cover, rather than cliffs that expose rocky outcrops and are actively receding. Some bluffs originated as subaerially-weathered slopes in Pleistocene phases of lowered sea level, and have persisted during and since the Late Quaternary marine transgression because the sea that rose alongside them was relatively calm, or because there was some protective feature, such as a coastal terrace, a wide beach or a coral reef, which prevented them being attacked and undercut by strong wave action. There are many steep forested slopes in the humid tropics, notably in the sheltered coasts of the Indonesian archipelago (Bird and Ongkosongo 1980) and in north-eastern Australia, where the Great Barrier Reef excludes strong oceanic wave action (Bird 1989a). These show occasional slumping of the weathered mantle, especially after the slope foot has been undercut by storm waves, but the scars are usually healed quickly by luxuriant vegetation.

Other bluffs were former sea cliffs that became degraded by subaerial weathering after marine erosion ceased, either because of emergence (due to land uplift or sea lowering), which exposed the earlier foreshore as a protective fringe so that waves no longer reached the cliff base, or because of the accumulation of sediment, usually sand or shingle beaches, sometimes with dunes, in the form of a protective foreland.

As sea level rises, deepening nearshore waters will submerge at least part of the existing intertidal zone, thereby intensifying wave attack at the base of cliffs and bluffs, and accelerating their erosion. Accelerated cliff recession has occurred during phases of rising water level around the Great Lakes, notably at Scarborough Bluff, near Toronto (Carter and Guy 1988). Bluffs will be undercut to form basal cliffs, and slumping will become more frequent on the vegetated slopes which will eventually develop into steep retreating cliffs (Fig. 22). Existing cliffs will generally become more unstable, and recede more rapidly, areas of land being lost as sea level rises (Fig. 23). The exception will be where rock outcrops are so resistant that the high and low tide lines simply move up the existing cliff face. Clayton (1989) estimated that in Britain, cliffs that are already retreating 1 m per year would show an accelerated retreat of 0.35 m per year for every 1 mm per year rise in sea level. On the volcanic island of Nii-Jima, off the Japanese coast south of Tokyo, Sunamura (1988) calculated that cliffs cut in poorly consolidated volcanic gravel and ash, now retreating at 1.2 m per year, will be cut back at up to 2.3–2.9 m per year, and will thus have receded 230–290 m by the year 2100.

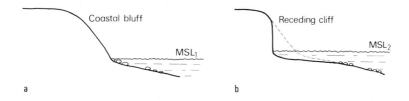

Figure 22   As sea level rises, coastal bluffs will become receding cliffs, and recession of existing cliffs will accelerate. (a) Present situation. (b) Response to sea level rise

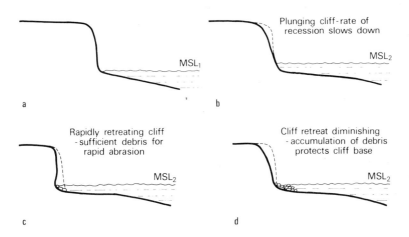

Figure 23   The effects of a sea level rise on a cliffed coast. (a) Present situation. (b)–(d) After sea level rise. If there is no beach material for abrasion (b), a plunging cliff will develop; if there is sufficient material for wave action to use as ammunition in attacking the cliff base (c), there will be rapid cliff recession; and if debris accumulates to protect the base of the cliff (d), cliff retreat will diminish

Where shore platforms have developed in front of cliffs or bluffs they will be submerged for longer periods, and become permanently inundated where the sea level rise exceeds the tidal range. Waves that now reach the base of the cliff only at high tide will be able to maintain a more consistent attack, thereby accelerating cliff erosion. Submerging shore platforms may acquire a veneer of sediment, or be mantled by accretionary growths of nearshore plant and animal communities. Cementation of sandy ripples on the surface of a shore platform cut in Pleistocene dune calcarenite on the coast near Sorrento, in south-eastern Australia, has been cited as a possible consequence of a slight relative sea level rise (Bird 1982), as have algal encrustations on formerly eroding shore platforms near Port Hedland in Western Australia, and at Nyali in Kenya (Bird and Guilcher 1982). Such changes may at least partly offset the effects of nearshore deepening and intensification of wave attack on backing cliffs.

Generally, cliff recession has been intermittent, cliff crests receding during episodes of storm erosion or slumping, then remaining in position, perhaps for many decades. On the chalk coasts of southern England and northern France cliff recession is mainly the outcome of rock falls, which occur particularly after the outcrops have been frozen during a cold winter and are then thawed. Talus cones of chalk thus formed are then gradually reduced by weathering and corrosion (the chalk being dissolved by rainwater and aerated spray and surf) and removed by wave erosion. The vertical cliff profile thus restored is then further undermined until it collapses again (May and Heeps 1985).

On the soft limestone cliffs of the Port Campbell coast in south-eastern Australia a sector 200 m long and up to 12 m wide collapsed into the sea in 1939, forming a basal talus (Baker 1943) that has been gradually reduced by marine erosion, but is still clearly visible—there has been no further cliff crest recession here in the ensuing half century. Evidence of such slumping is present on about 5% of the length of the Port Campbell cliffs, suggesting a recurrence interval of many decades for such major landslides. A rising sea will increase the frequency of these events, but it will be difficult to recognise an acceleration of cliff retreat, at least in the early stages, unless existing average rates of recession have been measured with sufficient accuracy to provide a basis for comparison. Such monitoring is available only for very limited parts of the world's cliffed coastline (Bird 1985).

On some cliffed coasts a basal notch has been formed close to high tide level by abrasion or solution. There are well-known examples on the limestone stacks of Phangna Bay in Thailand, and similar notches have been found at various levels above and below the present, notably on Pacific islands and around the Mediterranean, where they have been correlated with earlier stillstands of sea level (Pirazzoli 1986b). A rising sea will enlarge existing notches upwards, rather than develop a new notch at a higher level: the latter would form only after a rapid sea level rise was followed by a new stillstand.

On less resistant rocks, such as soft sandstones or clays, waves have shaped the nearshore sea floor into a concave profile that declines seaward from the cliff base, flattening below low tide. As cliffs recede, the nearshore sea floor tends to be lowered in such a way as to move this concave profile landward. This has been the case, for example, where cliffs and sea floor are cut into Pleistocene glacial drift deposits on rapidly receding sectors of the coast of eastern England (Cambers 1976). A sea level rise is likely to cause further landward migration of these concave sea floor profiles as the cliff base retreats.

Where cliffs of soft material have formed on coasts sheltered from strong wave action, they are usually dominated by gullying, slumping and other features resulting from subaerial weathering and the effects of runoff and seepage, rather than from marine processes. Removal of basal talus and downwashed deposits by wave action is then necessary if a vertical cliff profile is to be maintained. On such cliffs, a rising sea level is likely to increase marine erosion, reducing and eventually suppressing the features developed by subaerial processes, as undercutting of the cliff base produces a steeper or vertical cliff. The sequence will be similar to that documented from the gullied cliffs at Black Rock Point, on the shores of Port Phillip Bay, Australia (Bird and Rosengren 1987), which became steeper and smoother in profile after cliff-top stabilisation diverted runoff during heavy rains.

A sea level rise could initiate major landslides or produce new and more extensive slumping along coasts where the rock formations dip seaward, as on the coastline of Lyme Bay in southern England, where extensive mass movements have been taking place intermittently. The frequency of slumping would accelerate on a cliff cut back through horizontal or landward-dipping strata into an area where the rocks dip seaward.

Rates of cliff recession are influenced by the availability of rocky debris, including beach material, which can be mobilised by wave action and used as ammunition for cliff-base abrasion. Cliffs fringed by a narrow beach which is used in this way have retreated more rapidly than cliffs on the same formation where the beach is wide and high (so that waves do not reach the cliff base), or where there is no beach material (so that the cliff base is attacked only by the hydraulic action of breaking waves). If a rising sea submerges or disperses a narrow beach in such circumstances abrasion will diminish, and with purely hydraulic action cliff retreat could decelerate. If during a sea level rise sediment is eroded from a cliff or brought alongshore and deposited to form a persistent talus apron or beach in front of the cliff, wave action will be impeded and cliff retreat will slow down, and perhaps come to a halt. On the other hand, development of a narrow beach along the base of a cliff where there was previously a wide, protective beach (or no beach at all) will result in more vigorous abrasion and accelerated cliff retreat.

One approach to estimating rates of cliff retreat as sea level rises is to compare recession rates on similar rock formations with differing exposure. The retreat of a cliff fronted by an intertidal sloping shore platform, and subject to only brief episodes of basal wave attack at high tide, can be compared with a cliff cut in similar material where wave attack is stronger and more persistent because nearshore water is deeper, the latter situation simulating conditions after the sea has risen on the former. Such a contrast can be found where a longshore dip in strata of contrasted resistance produces some sectors where the cliff base is protected by a hard formation and intervening sectors where the intertidal zone coincides with a softer rock outcrop that has been scoured away. On the coast of Kimmeridge Bay, in southern England, the cliff is being cut back more rapidly behind sectors where the nearshore sea floor declines steeply than where the cliff base is protected by a flat structural ledge of limestone. After a sea level rise has submerged the structural ledge, the rate of cliff recession will increase, while that of the unprotected sectors will also accelerate as the bay deepens.

There have been few measurements of cliff recession rates on subsiding coastlines, but on the coast of Lyme Bay, in southern England, which is believed to have been subsiding at about 2.0 mm per year, surveys by Brunsden and Jones (1980) indicated that the slumping cliffs have been retreating at an average rate of 40 m per century. They did not determine how much of this retreat was due to subsidence, but suggested that if the sea level rise accelerated 'then either the sea cliff will move back faster than the mass movement processes can adjust, and a high rectilinear cliff will be formed, or the system will keep pace and maintain a similar, but even more dynamic, morphology'. The second possibility is more likely if sea level rise is accompanied by a climatic change that increases rainfall, seepage and runoff on this coast.

At Huntington, in southern California, land subsidence as the result of oilfield exploitation has led to a local, relative sea level rise during recent decades. Wave action intensified, accelerating the retreat of clay cliffs on the Huntington coast, until this was halted by extensive boulder armouring (Mayuga and Allen 1969).

BEACH AND BARRIER COASTS

Beaches, generally of sand or shingle (beach gravel), or mixtures of the two, are found on about one-third of the world's coastline. They occur where suitable sediment has been supplied to the coast by rivers or glaciers, or derived from the erosion of cliffs and rocky shores, and drifted along the shore, or where it has been washed in from the sea floor. Some sandy beaches have been supplied with wind-blown sand by dunes spilling from the hinterland, and in recent decades there has been artificial nourishment of selected beaches by the dumping of sand, gravel or other suitable rocky debris (Fig. 24).

Beaches also lose sediment offshore and alongshore, and into the mouths of estuaries and inlets, it can be washed landward over barriers and spits into lagoons and swamps, and sandy material can be blown from the beach by the wind. There is also a gradual attrition of beaches as a result of the diminution in grain size which occurs when sand and shingle are agitated and worn by wave action. Some beaches have been depleted by the removal of sand and gravel from the shore for use in road-making and construction industries or by the quarrying of mineral deposits (Fig. 24). Where the supply of sediment to a beach exceeds the losses the beach is built up and extended seaward, a process termed progradation. Where the losses are greater than the supply, beaches are depleted by erosion. Progradation has occurred where beach material drifting alongshore

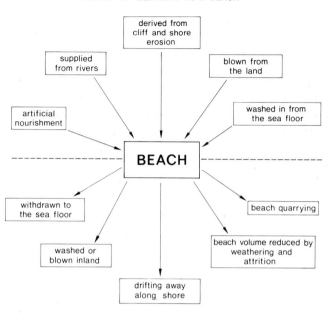

SUPPLY OF SEDIMENT TO A BEACH

derived from
cliff and shore
erosion

supplied
from rivers

blown from
the land

artificial
nourishment

washed in from
the sea floor

BEACH

withdrawn to
the sea floor

beach quarrying

washed or
blown inland

beach volume reduced by
weathering and
attrition

drifting away
along shore

LOSSES OF SEDIMENT FROM A BEACH

Figure 24   The sources of supply of sediment to a beach, and the ways in which sediment is lost from a beach

has been trapped alongside breakwaters or headlands, or has accumulated as spits and forelands, or within embayments.

Beaches shaped by longshore drifting, due to the oblique arrival of waves and to longshore currents, develop 'drift alignments', whereas those occupying compartments between headlands have 'swash alignments' (Davies 1972), determined by wave crests that are refracted so that they anticipate and, on arrival, fit the beach outline in plan. The former can be termed 'drift-dominated beaches', which often show irregular migrating lobes and forelands and culminate in longshore spit growth; the latter are 'swash-dominated beaches', usually smoothly curved, sometimes asymmetrical in plan. Some beaches are composite, dominated by longshore drifting when waves arrive obliquely, and becoming swash-dominated when waves arrive to fit the shoreline. The Ninety Mile Beach, in south-eastern Australia, is largely swash-dominated because of the prevalence of refracted south-easterly ocean swell, but winds from the south-west and the east can generate waves that produce longshore drifting to the north-east and south-west, respectively (Bird 1984).

Many drift-dominated beaches are narrow, fringing cliffed coasts or the shores of deltas or coastal plains, and continuing alongshore as spit formations, but on some coasts beach material has accumulated in such a way as to form a beach-ridge plain, with beaches backed by low parallel sand or shingle beach ridges. These result from alternating progradation and erosion (a process of 'cut-and-fill', described below), where there has been a sustained sediment supply to a coast that has been tectonically stable or gradually emerging. Sometimes the beach ridges are surmounted by dune ridges, and in places they are interspersed with lagoons and swamps. Where the supply of sand has been abundant and onshore winds are strong the beach ridges may be overrun by large landward-spilling dunes.

On some coasts, progradation has led to the formation of a barrier or chain of barrier islands in front of lagoons and swampy areas (Schwartz 1973). On the global scale, these are built mainly of sand, but some (such as Chesil Beach on the south coast of England) consist of shingle, and others are mixtures of sand and shingle. Some barriers originated as spits that grew along the coast to enclose lagoons; others developed as a result of the accumulation of sediment moved shoreward from the sea floor; and some have formed by a combination of longshore and onshore sediment supply.

Many coastal barriers have wave-built beach ridges or wind-built foredunes which are 'lines of growth' marking successive coastlines, and indicating intermittent progradation, with phases of erosion and truncation, as on Galveston Island in Texas and St Vincent Island in Florida: on the Pacific coast of Mexico the Nayarit barrier enclosing Laguna Agua Brava has prograded to form a beach-ridge plain up to 15 km wide. Such barriers are anchored in position, parallel to the pre-existing coastline, and have been widened by progradation, as on the Gippsland coast in south-east Australia (Bird 1984).

Other coastal barriers and barrier islands are transgressive, as on parts of the Gulf and Atlantic coasts of the United States. They have been built upward and driven landward during the Holocene (Kraft, Biggs and Halsey 1973), and some are still moving intermittently landward as the outcome of storm overwash and the drifting of dunes driven by onshore winds (Leatherman 1981). In the south-eastern North Sea, Hofstede (1991) has traced the coastward migration of sand islands such as Scharhörn in recent decades as a consequence of increasingly frequent storm surges in an area where sea level is rising because of land subsidence. The shingle barrier of Chesil Beach in southern England developed and was driven landward by recurrent storm surges in such a way as to prevent waves from the open sea reaching the mainland coast, so that the gentle slopes of the embayed shoreline behind the Fleet lagoon lack marine cliffing (Steers 1953). In general, transgressive barriers are best developed on coasts that are subsiding, whereas anchored and prograded barriers are generally associated with stable or emerging coastlines.

Most beaches show short-term (<1 year) alternations known as 'cut and fill', with erosion (cut) when some of their sediment is withdrawn seaward during episodes of storm wave activity, and deposition (fill) in intervening calmer periods when gentler wave action moves sediment back shoreward to be deposited on the beach face. Cycles of erosion and deposition may also result from minor oscillations of sea level, as demonstrated on the shores of Lake Michigan (Fig. 25) by Olson (1958). A beach system may be said to be 'in equilibrium' when the transverse profile of the beach and adjacent nearshore zone alternates cyclically, in response to processes generated by wave action in nearshore waters, without long-term (>5 year) changes; an equilibrium which is lost on beaches that continue to advance by deposition (prograding) or retreat because of erosion.

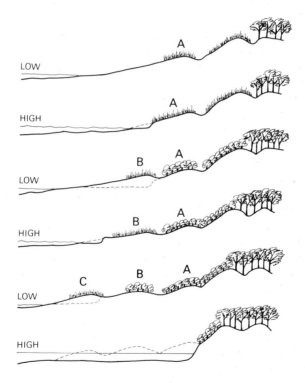

Figure 25   Sequence of changes on a sandy shore bordering Lake Michigan, North America, as a result of successive oscillations of water level, based on a diagram by Olson (1958). Beach ridges were built up as dunes behind the prograding beach during episodes of falling lake level, then trimmed back, and eventually destroyed, by erosion during phases of rising lake level

Various kinds of nearshore topography accompany beaches. Some descend to rocky or muddy sea floors that may be partly exposed at low tide; others pass down into smooth, usually concave, nearshore profiles of sand or shingle; and others are fronted by depositional bar formations which run parallel to the coastline, or occur in more intricate patterns. These are usually of sand, occasionally gravel, and studies of their morphology have shown that they are dynamic features, shaped and moved largely by storm wave activity. Where sand banks move along the coast there may be beach accretion as they arrive and erosion as they move away.

A world-wide study of coastline changes by the International Geographical Union's Commission on the Coastal Environment (IGU-CCE) between 1972 and 1984 demonstrated a global prevalence of beach erosion (Bird 1985). It found that in recent decades there has been erosion on more than 70% of the world's sandy coastlines, less than 10% having prograded, while the remaining 20%–30% have either remained stable or been subject to alternations with no net change.

The several reasons for the onset of erosion on beaches that were previously stable or prograding are listed in Table 1. Beach erosion has generally resulted from a combination of several factors, one or two of which may have been dominant (Bird 1987). On some coasts it is because the supply of sediment from rivers has diminished, largely because of the building of dams that have intercepted sediment upstream in reservoirs. On others, there has been a reduction in sand and shingle supply from alongshore, especially where cliffs have been stabilised and are no longer a sediment source, or where breakwaters have been built, and have intercepted beach material drifting along the coast. In some cases there has been a reduction in sediment supply from the sea floor. Certain beaches have been depleted by quarrying for sand and gravel, or for minerals such as rutile, tin or gold. There are places where beach erosion has followed increased wave attack because of the deepening of nearshore water, either by dredging or as a result of subsidence. Short-term sea level rises of the kind that occurred along the Pacific coast of North America, during the 1982–1983 El Niño Southern Oscillation (ENSO), when sea level rose 20–30 cm, have also generated beach erosion (Komar and Enfield 1987). Bryant (1983, 1985, 1988) found that phases of beach erosion at Stanwell Park, New South Wales, were correlated with episodes of rising sea level and heavier rainfall accompanying the ENSO. An increase in storminess in coastal waters can also result in the erosion of beaches that were previously stable or prograding.

There is no doubt that a global sea level rise will initiate beach erosion, or accelerate it where it is already taking place, but it has not been the primary cause of the widespread beach erosion that has already occurred during the past century (Fig. 26). Beaches have been eroding on coasts where no sea level rise has been registered, and even on emerging coasts, where the land has risen

Table 1   The causes of beach erosion

1. Submergence and increased wave attack as the result of a rise in sea level or as the result of coastal land subsidence: the 'Bruun Rule'.
2. Diminution of fluvial sand and shingle supply to the coast as a result of reduced runoff or sediment yield from a river catchment (e.g. because of a lower rainfall, or dam construction leading to sand entrapment in reservoirs, or successful revegetation and soil conservation works).
3. Reduction in sand and shingle supply from eroding cliffs or shore outcrops (e.g. because of diminished runoff, a decline in the strength and frequency of wave attack, or the building of sea walls to halt cliff recession).
4. Diminution of sand and shingle supply washed in by waves and currents from the adjacent sea floor, either because the supply has run out or because the transverse profile has attained a concave form which no longer permits such shoreward drifting.
5. Reduction in sand and shingle supply from the sea floor because of increased growth of seagrasses or other marine vegetation, which impedes shoreward drifting.
6. Diminished production of sand and shelly deposits from sea floor biogenic sources because of ecological changes reducing the production of shelly material.
7. Reduction in sand and shingle supply from alongshore sources as the result of interception (e.g. by a constructed breakwater).
8. Increased losses of sand from the beach to the backshore and hinterland areas by landward drifting of dunes, notably where backshore dunes have lost their retaining vegetation cover and drifted inland, lowering the terrain immediately behind the beach and thus reducing the volume of sand to be removed to achieve coastline recession.
9. Removal of sand and shingle from the beach by quarrying or the extraction of mineral deposits.
10. Reduction of sand supply to the shore where dunes that had been moving from inland are stabilised, either by natural vegetation colonisation or by conservation works, or where the sand supply from this source has run out.
11. Losses of sand and shingle from intensively-used recreational beaches.
12. Increased wave energy reaching the shore because of the deepening of nearshore water (e.g. where a shoal has drifted away, where seagrass vegetation has disappeared, or where dredging has taken place).
13. Increased wave attack due to a climatic change that has produced a higher frequency, duration or severity of storms in coastal waters.
14. Diminution in the calibre of beach and nearshore material as a result of attrition of beach sand grains, leading to winnowing and losses of increasingly fine sediment from the shore.
15. Diminution in the volume of beach and nearshore material as a result of weathering, solution or attrition, resulting in the lowering of the beach face and a consequent increase in penetration of wave attack to the backshore.
16. A rise in the water table within the beach, due to increased rainfall or local drainage modification, rendering the beach sand wet and more readily eroded.
17. Increased losses of sand and shingle alongshore as a result of a change in the angle of incidence of waves (e.g. as a result of the growth or removal of a shoal or reef, or breakwater construction).
18. Intensification of wave attack as a result of lowering of the beach face on an adjacent sector (e.g. as the result of reflection scour induced by sea wall construction).
19. Migration of beach lobes or forelands as a result of longshore drifting—progradation as these features arrive at a point on the beach is followed by erosion as they move away downdrift.
20. On arctic coast beaches the removal of a protective sea ice fringe by melting, so that waves reach the beach (e.g. for a longer summer period).

Figure 26   Beach erosion is extensive around the world's coastline. Here, on the southern coast of Jutland, Denmark, erosion has cut the beach back sufficiently to undermine defensive blockhouses built during the German occupation in the 1940s

relative to the sea: on the Indalsälven delta in north-east Sweden, beach erosion has been due primarily to a recent reduction in fluvial sediment yield following dam construction upriver (Bird 1985). Nevertheless, a global sea level rise will undoubtedly result in beach erosion becoming even more extensive and severe than it is now because submergence will deepen nearshore water, so that larger waves break upon the shore (Fig. 27). Where beach-ridge plains and coastal barriers have been formed as a result of Holocene progradation, a sea level rise is likely to initiate or accelerate erosion along their seaward margins (Fig. 28). Beaches already in retreat will be cut back more rapidly, but they will continue to exist as the coastline retreats through beach-ridge plains or coastal barriers. In some areas, the sand supply may be maintained, or even increased, as a result of accelerated cliff erosion, or greater sediment yield from rivers because of heavier rainfall and stronger runoff, catchment devegetation, or disturbance by tectonic uplift or volcanic activity. If the nearshore sea floor is built up by sediment accretion in such a way as to maintain or aggrade the existing transverse profile, or if longshore drifting supplies sufficient sediment, beaches may be maintained, or even prograded, during a sea level rise.

Beaches will disappear from many sectors where they are already narrow, and backed by high ground, unless the sea level rise increases cliff erosion and

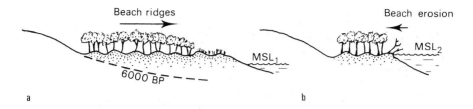

Figure 27   Beaches backed by parallel beach ridges formed by progradation during the Holocene will be cut back by erosion as sea level rises, except where there is a sufficient sediment supply to the coast to offset such erosion. (a) Present features. (b) After sea level rise

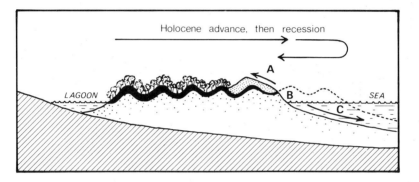

Figure 28   The evolution of many barrier formations in Holocene times included progradation, with development of successive beach or dune ridges. This has often been followed by recession of the seaward margin, with losses of sand landward into spilling dunes (A), alongshore (B), and seaward to the sea floor (C). The sequence generally results from a change from an abundant supply of sand during the early Holocene to a deficit during the late Holocene. A sea level rise will intensify this erosion as larger waves move in through deepening nearshore water

generates additional sediment to maintain them. Beaches that front salt marshes or mangrove swamps are likely to be eroded and overwashed, with the sea beginning to attack the formations that lie behind them. Where sea walls have been built to halt coastline recession, beaches fronting them may be depleted or removed altogether by scour due to reflection of incident waves. On the coast of Britain, numerous groynes have been built in attempts to intercept and retain beach sediment drifting alongshore. Some have succeeded, but often at the cost of impoverishing beaches and causing erosion farther along the coast.

Coastal barriers that are already transgressive will continue to migrate landward as sea level rises, and some stationary barriers that prograded during the Holocene stillstand may become transgressive as a result of erosion along their seaward margins, with overwash and landward drifting of dunes. It is of interest that on the very gently sloping (gradient about 1:1000) sandy areas bordering the south-west Caspian a 1.5-m sea level rise during the past decade has led to the formation of transgressive barriers in front of shallow lagoons. This barrier–lagoon assemblage is now migrating landward as the sea rises (Leontiev and Veliev 1990). As beach erosion cuts into backshore dunes, blowouts are initiated, and these can grow into large transgressive dunes as sand is excavated and blown landward. A rising sea level will accentuate these effects (Carter 1991, Van der Muelen 1990). The lowering of the backshore topography by such landward losses may result in accelerated beach erosion, because a smaller volume of material has to be consumed to achieve a specific length of coastline recession. Similar acceleration will occur where backshore terrain has been lowered by sand mining. If the climate becomes drier, a reduced vegetation cover may allow backshore dunes to be blown away, whereas if it becomes wetter, enriched vegetation may hold them in place, so that the backshore is not lowered.

Where subsidence is in progress, as on the Atlantic seaboard of the United States, sea level rise has contributed to beach erosion. Bruun (1962) proposed a model of the response of a sandy beach to sea level rise in a situation where the beach was initially in equilibrium, neither gaining nor losing sand. Erosion of the upper beach would then occur, with removal of sand to the nearshore zone in such a way as to restore the previous transverse profile (Fig. 29). In effect, there would be an upward and landward migration of the transverse profile, so that the coastline would recede beyond the limits of submergence. This restoration would be completed when the sea became stable at a higher level, and coastline recession would come to an end after a new equilibrium was achieved. The extent of recession was predicted by using a formula that translates into a 'rule of thumb' whereby the coastline retreats 50–100 times the dimensions of the rise in sea level: a 1-m rise would cause the beach to retreat by 50–100 m. Since many seaside resort beaches are no more than 30 m wide, the implication is that these beaches will have disappeared by the time the sea has risen 15–30 cm (i.e. by the year 2030), unless they are artificially replaced.

This model (known as the Bruun Rule) has been supported by laboratory experiments, and by the changes that have occurred, for example around the shores of the Great Lakes during and after episodes of rising water level (Schwartz 1965, 1967). If there are no gains or losses from the hinterland or alongshore it is possible to estimate how much retreat will occur, but in practice there are problems in using the Bruun Rule to forecast the extent of coastline recession in response to a predicted sea level rise.

The first problem is to determine seaward boundary conditions for application of the Bruun Rule (i.e. the extent offshore of the profile that is to be restored at a higher level). Bruun (1988) suggested that the boundary should be the

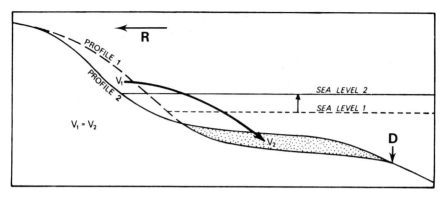

Figure 29   The Bruun Rule states that a sea level rise will lead to erosion of the beach and removal of a volume of sand (v1) seaward to be deposited (v2) in such a way as to restore the initial transverse profile landward of D, the outer boundary of nearshore sand deposits. The coastline will retreat (R) until stability is restored after the sea level rise comes to an end. The coastline thus recedes further than it would if submergence were not accompanied by erosion

limiting depth between predominant (coarser) nearshore and (finer) offshore material, where the water is too deep for waves to move material of beach calibre, but this requires that detailed sedimentological surveys of the sea floor be available before the sea rise begins. It is not always possible to distinguish nearshore from offshore material: in many places there are gradual transitions, or nearshore topography is variable, with sand bars, rocky outcrops, or a muddy substrate immediately offshore.

Several attempts to test the Bruun Rule have been reviewed by the Scientific Committee on Ocean Research (SCOR) (SCOR Working Group 1991). Rosen (1978) found good agreement between actual retreat on the submerging shores of Chesapeake Bay and the extent of recession that would have been predicted by the Bruun Rule, but on the shores of the Great Lakes Hands (1983) found that although beach profiles began to adjust according to the Bruun Rule as lake level rose 1 m, the adjustment was incomplete when lake level began to fall again. The Rule requires a sea level rise to be followed by a period of stillstand when the adjustment to the changed conditions can be completed. It may give misleading results if applied to the predicted scenario of an accelerating sea level rise.

On the generally low to moderate wave energy coasts of New Jersey and Maryland, Everts (1985) found that the Bruun Rule overestimated beach recession, but on the Pacific coast, where wave energy is greater, erosion was between two and four times as much as predicted.

Where the transverse gradient is very low, as on parts of the Caspian coast (Kaplin 1989), application of the Bruun Rule is complicated by the fact that waves re-shape the upper beach profile, throwing some sand up to form a barrier which is then driven landward as sea level continues to rise. As it does so, the steepened beach on the seaward side maintains its profile partly by landward overwash and partly by seaward transference of sand. The results follow the model developed by Dean and Maurmeyer (1983) for the response of a barrier island to a phase of rising sea level, incorporating aggradation and overwash as well as adjustment of the seaward shore. These examples indicate that it would be preferable to adapt the Bruun Rule in terms of sediment budgets rather than two-dimensional profile responses.

A precondition of the Bruun Rule as originally proposed was that the beaches subjected to a sea level rise were initially in equilibrium. As has been noted, beach erosion is already widespread, and only a small proportion of the world's sandy beaches can be considered to be in equilibrium. The Rule must also be qualified to deal with any gains or losses that occur as a result of longshore drifting and landward movement of sand from beaches into tidal inlets or on to, and across, barriers or spits into backing lagoons and swamps. It takes no account of changes in processes or rates of sediment supply and removal due to climatic variations accompanying a sea level rise, such as increased rainfall and runoff, stronger winds and stormier seas. It should be emphasised that the Bruun Rule covers only one of the twenty factors identified in Table 1 as causes of beach erosion.

Fisher (1984) found that between 1939 and 1975 the barrier beach coast of Rhode Island retreated at an average rate of 0.2 m per year in a period when sea level rise averaged 0.3 cm per year. This was within the range predicted by the Bruun Rule for this rate of sea level rise: 0.15–0.3 m per year, but Fisher calculated that 35% of linear beach recession over this period was due to losses of sand into tidal inlets and 26% to losses over the barrier islands to migrating dunes and washover fans. This left only 15% of the beach retreat accountable through submergence and 24% transference of beach sand seaward, as predicted by the Bruun Rule, which thus overestimated the response here (Fig. 30). The movement of sand over and between these barrier islands indicates that they are continuing the long-term landward migration, accompanied by the landward movement of lagoons and marshes, that has characterised much of the Gulf and Atlantic coastline during the Holocene. Beach ridge and barrier terrain that is not transgressive is likely to be cut back along the seaward margin, with losses mainly seaward and alongshore, as indicated by Teh Tiong Sa (1992) with reference to the sandy beach ridge terrain known as permatang on the coast of Malaysia.

Where beaches are bordered by nearshore sand bars a sea level rise may be accompanied by upward growth and landward movement of the sand bars as the beach is cut back, providing there is a sufficient supply of suitable sediment available to maintain this landform association. The implication is

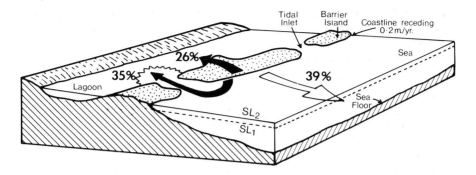

Figure 30 Changes on the sandy barrier island coast of Rhode Island, USA between 1939 and 1975, based on calculations reported by Fisher (1984). He found that over this period mean sea level rose 30 cm/year. The average annual rate of coastline retreat was 20 cm/year, of which 35% was due to losses into tidal inlets, 26% to the washing and blowing of sand landward across the barrier islands, and 39% to submergence and sand movement from the beach to the sea floor. (a) Present situation. (b) Response to sea level rise

that the landward migration of the outer limit of the sand bars would be equal to the extent of beach recession, and that the overall profile would be preserved. A complication is that sand bars usually consist of finer material than is present on the beach face, so that erosion and seaward movement of beach sand may not provide sediment of suitable calibre for their maintenance, at least in their existing form. On the shores of Lake Michigan, Dubois (1977) found that beach face erosion as lake level rose was matched by accretion on the landward side of the nearshore bar, which thus widened, its seaward side remaining unchanged.

On The Netherlands coast, Edelman (1972) modelled the gains that occurred on beaches backed by dunes during calm weather phases and the losses when storm surges temporarily raised sea level. The losses have outweighed the gains. De Ronde (1988) calculated that over a 20-year period 5.3 million m^3 of sand were lost from eroding sectors of the Dutch coast, and only 3.4 million m^3 gained on accreting sectors. The deficit of 1.9 million m^3 has been partly withdrawn to the floor of the North Sea, and partly washed in through tidal entrances in the West Frisian barrier island chain to be deposited in the Wadden Sea.

Prediction of the extent of beach recession as sea level rises thus faces a number of difficulties. As long as the sea is rising, beach-fringed coastlines will continue to recede as erosion accompanies submergence, and if the sea rises at an increasing rate the beach erosion will accelerate. It is not possible to predict the precise location of a sandy coastline after sea level has risen 1 m when it is continuing to rise: all that can be said with this scenario is that in

100 years' time most sandy coastlines will have retreated substantially, and that (in the absence of protective structures or artificial nourishment) they will be eroding more rapidly. A more precise forecast would be possible if a phase of sea level rise were followed by a stillstand, which could occur after the human modifications of the atmosphere that enhance the greenhouse effect were brought under control.

ESTUARIES AND COASTAL LAGOONS

During the world-wide Late Quaternary marine transgression valley mouths were inundated to form inlets (estuaries) into which rivers flowed, and coastal lowlands were submerged to form broad embayments. The inlets at river mouths have been modified during the ensuing 6000 years of relative stillstand by sedimentation from the rivers and from the sea, which has built up bordering alluvial land and produced shoals with intervening channels. The waters of estuaries are generally brackish as a result of interactions between marine tidal incursions and fresh water from rivers.

In some cases the inlets have been filled in completely to form alluvial valley floors, which eventually protrude seaward as deltas. Alternatively, spits and barriers of sand or shingle have been built up across the mouths of inlets and embayments in such a way as to enclose lagoons. These are generally shallow, and linked to the open sea by one or more marine (tidal) entrances; they are typically estuarine, with a salinity gradient increasing from the mouths of inflowing rivers to sea water at tidal entrances. An example has been described from the Gippsland Lakes in south-eastern Australia (Bird 1978). Shallow areas behind chains of barrier islands (such as the Wadden Sea behind the Dutch, German and Danish Frisian Islands) can be considered 'open lagoons', where the ratio of tidal entrance width to longshore barrier length exceeds 1:10.

Coastal lagoons have been reduced in depth and area by sedimentation from inflowing rivers, as well as by accumulation of sediment washed in from the sea, organic deposits such as peat and shells, and precipitated material, notably carbonates and chlorides. Swamp encroachment during the past 6000 years has resulted in shallowing and shrinkage of many lagoons. Changes in configuration are related to the effects of wind-generated waves and the currents produced by rivers, winds and tides within a lagoon.

Ecological conditions, notably water salinity, have been important in controlling the extent to which vegetation (e.g. reedswamp, salt marsh, mangroves) can colonise estuary and lagoon shores, impeding erosion, promoting patterns of sedimentation, and generating organic deposits that are usually incorporated in accreting substrates. A salinity increase may result from diminished runoff from the hinterland, as where river outflow has been intercepted by dam construction or because of greater salt accessions from the subsoil, following vegetation clearance, augmented rainfall, or an artificially raised water table (e.g. by irrigation of adjacent farmland). Alternatively, lagoon salinity may increase after the opening or enlarging of a marine (tidal) entrance, as in the Gippsland Lakes, Australia (Bird 1978), where the increased salinity has led to changes in configuration as a result of losses of shoreline vegetation and consequent erosion, with salt marshes replacing reedswamp. In Louisiana increasing salinity has killed cypress swamps and converted them into brackish lagoons (De Sylva 1986), while on the Egyptian coast diminishing supplies of fresh water from the River Nile have resulted in an increase in salinity in coastal lagoons such as Lake Burullus, and a consequent decline in the sardine fishery.

Reduction of lagoon salinity by an increase in rainfall and freshwater inflow has the opposite effect, stimulating freshwater vegetation and swamp encroachment. This occurred after the completion of barrages built to exclude sea water from the Murray mouth lagoons in South Australia in 1940, bordering salt marshes having since been replaced by encroaching reedswamp shores (Bird 1962). Similar changes followed the diversion of the Waitangitaona River into formerly brackish Okarito Lagoon on the west coast of New Zealand's South Island during a flood in 1967.

As sea level rises, estuaries will tend to widen and deepen. Tides will penetrate farther upstream, tide ranges may increase, and there will be changes in patterns of shoal deposition. The discharge of river floods will be impeded as sea level rises, so that flooding becomes more extensive and persistent, and a higher proportion of fluvial sediment will be retained within the submerging estuaries, instead of being delivered to the sea floor or adjacent coasts. There will be increasing penetration upriver by salt water, which may also invade underground aquifers in coastal regions, especially if these have been depleted by groundwater extraction. Some of the effects will simulate those now briefly experienced during droughts: for example, in the 1960 drought, salt water spread up the Delaware River as far as Philadelphia. They will persist, however, with a higher sea level, and any drought effects then will lead to still further salt penetration. However, if there are accompanying climatic changes that increase rainfall in coastal regions or in the hinterland, some of these changes may be at least partly offset by greater fluvial discharge.

Coastal lagoons will generally be enlarged and deepened as sea level rises (Fig. 31), with submergence and erosion of their shores and fringing swamp areas, and widening and deepening of tidal entrances, increasing the inflow of sea water during rising tides and drought periods. Erosion of the enclosing barriers may lead to breaching of new lagoon entrances, and continuing erosion and submergence may eventually remove the enclosing barriers and reopen the lagoons as marine inlets and embayments. On the other hand, new lagoons may be formed by sea water incursion into low-lying areas on coastal plains, or

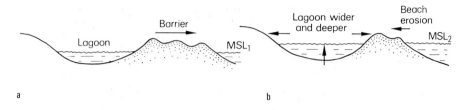

Figure 31   Coastal lagoons enclosed from the sea by barrier formations will widen and deepen as sea level rises, the barriers being narrowed, and possibly breached, by erosion along their seaward shores

where such depressions are flooded as the water table rises, forming seasonal or permanent lakes and swamps.

As sea level rises the currents that flow through existing tidal entrances and gaps between barrier islands, as on the Dutch and German North Sea coasts, may be strengthened, augmenting the inflow of water and sediment. In parts of the 'open lagoon' of the Wadden Sea, increased sediment inflow has been building up intertidal and nearshore mudflats and sandflats and maintaining vertical accretion on salt marshes on a coast that is subsiding.

There will be much variation in the nature and extent of changes in coastal lagoons as sea level rises, depending on their existing configuration and dynamics, and the extent to which they have been modified by human activities. The Venice Lagoon (Fig. 32), for example, has been maintained partly by continuing subsidence in the north-west Adriatic region, and partly by the diversion of rivers such as the Po (Fabbri 1985) and Brenta, which had been carrying sediment into it. Originally the aim was to maintain the lagoon as a natural defensive moat for the city and republic of Venice (Zanda 1991). In the 19th century industrial development led to the growth of Porto Marghera, near Mestre, and the need to provide ship access through the inlets in the enclosing barrier and across the lagoon to the petrochemical works established in 1972 resulted in the dredging of deeper approach channels. There was also extensive reclamation of bordering swamp areas, partly for the construction of fishponds known as valli (Gatto and Carbognin 1981). In recent decades the intertidal areas, known as barene, including salt marshes, have been shrinking as the result of submergence and erosion (Cavazzoni 1983). In addition, storm surge flooding in the city of Venice has become more frequent and severe in recent decades (Pirazzoli 1987). All these factors will influence the way in which the Venice Lagoon responds to an accelerating sea level rise.

The deepening and enlargement of estuaries and coastal lagoons may be countered where there is a supply of sediment, arriving at a sufficient rate to offset the effects of submergence. The Holocene stratigraphy of the New Jersey

63

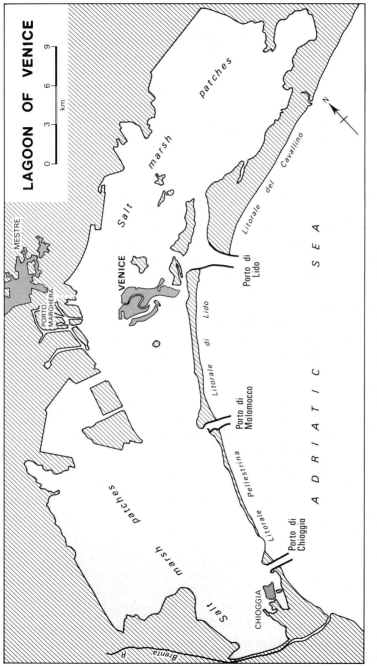

Figure 32   The lagoon of Venice

coast shows that sea level rise (due to land subsidence) during the past 2500 years was accompanied by the upward growth and expansion of salt marshes so that coastal lagoons backing sandy barrier islands were reduced in area (Psuty 1986). In southern Java the Segara Anakan (Fig. 33) is a rapidly shrinking coastal lagoon bordered by mangroves spreading forward on to tidal mudflats that are accreting as a result of sediment inflow from the Citanduy River, augmented in recent decades by soil erosion, due to the deforestation of steep headwater regions (Bird and Ongkosongo 1980). This rapid sedimentation will prevent much enlargement of this lagoon as sea level rises, but submergence will help to postpone the infilling of the residual lagoon basin.

The water of estuaries and coastal lagoons will become more saline as the sea rises because of increased marine penetration, and there is likely to be an increase in tide range and tidal ventilation. There will be accompanying ecological changes as areas that were previously fresh or slightly brackish become more saline, with die-back of freshwater vegetation and its replacement by halophytes, and changes in the distribution of fauna, with increases in marine and estuarine species as freshwater communities are displaced upstream. Freshwater fisheries will be reduced, and replaced by brackish and marine species. As has been noted, changes of this kind have already occurred in some lagoons for other reasons. In lagoons such as the Gippsland Lakes, already modified by the cutting of an artificial entrance, a sea level rise will accentuate ecological changes that have already been taking place.

Coastal lagoons show variations related to climatic conditions, especially the water balance determined by rainfall and river inflow on the one hand, and evaporation and sea inflow on the other. In arid regions the salinity of lagoons and semi-enclosed embayments is usually higher than that of the open sea. Some lagoons have dried out as evaporite plains, such as the Coorong in South Australia. In such systems a sea level rise may actually reduce salinity by forming a wider and deeper marine entrance, and increasing marine flushing. However, the water balance may also be modified by accompanying changes in climate, such as increasing rainfall.

DELTAIC COASTS

Deltas have formed where sediment brought down by rivers has filled the mouths of valleys that were drowned by the Late Quaternary marine transgression, and has been deposited to form low-lying land protruding into the sea. They have been built above normal high tide level by sedimentation from recurrent river floods, and wave deposition along their seaward margins. Their shaping is the outcome of interactions between supply of fluvial sediment (with which some material drifting alongshore, or in from the sea floor, may be incorporated) and

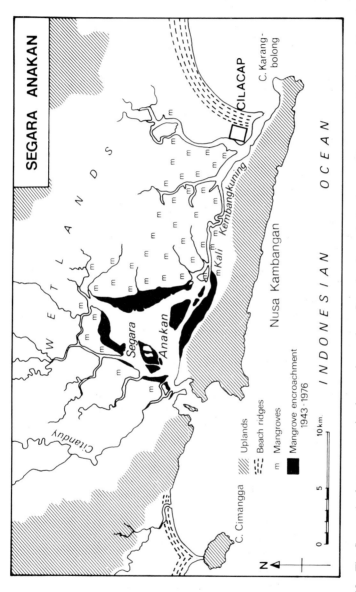

Figure 33 The Segara Anakan, an estuarine lagoon in southern Java, has been shallowing rapidly as a result of inwashing of sediment from rivers, notably the Citanduy. In recent decades there has been rapid mangrove encroachment. A rising sea level will here slow down mangrove encroachment and delay the infilling of the lagoon by fluvial sedimentation

the effects of marine waves and currents, which erode or build up the delta shoreline and disperse some of the sediment delivered to river mouths.

A delta continues to grow where sediment is accumulating at and around a river mouth above normal high tide level in excess of that carried away by waves and currents, and at a rate which more than offsets any subsidence in progress. Fluvial sedimentation is still prograding deltas, especially in the humid tropics, where broad swampy lowlands fringe deltas built where large rivers have delivered vast amounts of silt and clay to prograde the coast.

Fluvial sediment yields are greatest from rivers that drain lofty hinterlands and catchments with soft, erodible or deeply weathered geological outcrops, especially where human activities have resulted in deforestation, intensive grazing and cultivation, accelerating erosion on steep slopes. The Mediterranean region has had a long history of catchment deforestation and consequent soil erosion, leading to increased river sediment yields and the augmented growth of deltas such as those of the Po and the Ebro (Jelgersma 1988). On the north coast of Java, deposition around river mouths during episodes of flooding has been advancing the coastline locally by up to 200 m per year, partly as a result of an augmented sediment yield following deforestation and the development of agriculture in the steep hinterland. In China the Hwang Ho (Yellow River) is an example of a heavily laden river that drains a catchment of loess deposits undergoing rapid erosion as a result of vegetation clearance and intensive cultivation (Chen Jiyu, Lu Cangzi and Yu Zhiying 1985). It carries more than 1.2 billion tons of silt downstream to the coast annually, and its modern delta is growing rapidly out into Bo Hai Bay. Since its diversion to this outlet in 1855 it has added more than 2300 km^2 of land to the coast, and it is still growing seaward 2–3 km per year.

Fluvial sediment yields have also increased in rivers draining catchments where mining has been extensive, as in the Minas Gerais mining area in Brazil, on the south coast of Bougainville, and on the nickel-rich island of New Caledonia. Progradation has resulted at and around the mouths of these rivers.

Patterns of erosion and accretion on deltaic coastlines have been modified by natural or artificial changes in the positions of river mouths. Mention has been made of the Hwang Ho, which built its modern delta after it diverted northward into Bo Hai Bay, on the northern side of the Shandong Peninsula, during a major flood in 1855. For more than seven centuries the Hwang Ho had flowed into Jiaozhou Bay, on the Yellow Sea, and built a delta which has been cut back by wave action after the river changed its course during the 1855 floods (Chen Jiyu, Lu Cangzi and Yu Zhiying 1985).

Other deltas have had a history of changing positions of distributary river mouths, with alluvial (sub-delta) lobes developing around each mouth, then being removed by erosion after the mouth shifts. On the Danube delta the

Kilia lobe is still growing, while distributaries to older adjacent sub-deltas have waned. As the sub-deltas were dissected, silt and clay were dispersed, but sandy sediment was reworked and spread alongshore to form beaches and spits. The Rhône delta has had a similar history, indicated by truncated patterns of beach ridges around former river mouths, and the Mississippi delta has also been built in several stages, each initiated by a shift in the position of its outlet to the Gulf of Mexico (Fig. 34). In Java an old delta of the Cimanuk River has been removed by erosion since the river changed its outlet in a flood in 1947, and a new delta has grown out northward (Fig. 35). Similar changes took place after the nearby Solo River was diverted by cutting a canal to a new outlet, where the modern Solo delta is growing, and there has been rapid erosion of the abandoned delta (Bird and Ongkosongo 1980).

There has been erosion on deltaic coasts where fluvial sediment yield has been reduced as a result of the building of dams that impound sediment in reservoirs upstream. A well-known example is the Nile, where a succession of barrages built since 1902 culminated in the completion of the Aswan High Dam in 1964.

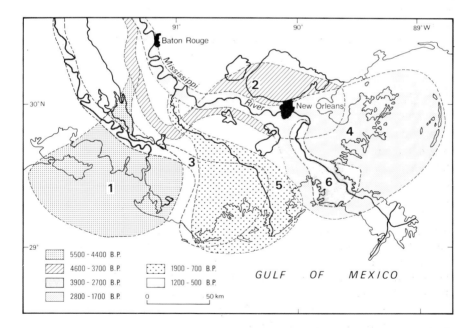

Figure 34 Stages in the evolution of the Mississippi delta, marked by the successive development of sub-deltas on a subsiding coast. The modern delta, south-east of New Orleans, is still growing, but the abandoned sub-delta lobes have become submerging coastlines. Redrawn from a diagram by Kolb and Van Lopik (1966)

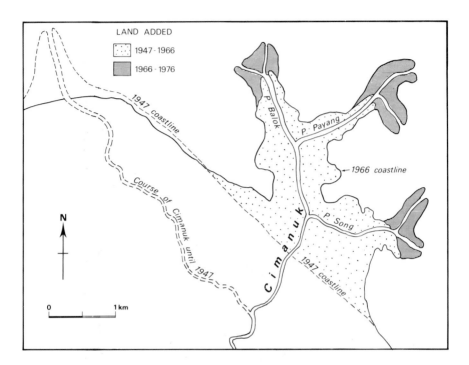

Figure 35   Growth of the Cimanuk delta, northern Java, seaward after its outlet was diverted during a flood in 1947

Subsequent interception of Nile sediment in the Lake Nasser reservoir, behind this dam, has been followed by the cessation of sediment flow downstream and the onset of rapid erosion, attaining more than 100 m per year locally near the mouths of the Rosetta and Damietta distributaries (Fig. 36) (Orlova and Zenkovich 1974, Sestini 1991). Dam construction upriver has also halted the growth of the Ebro delta, the shores of which are now eroding. On the Citarum delta in northern Java the onset of coastline erosion followed the completion of the Jatiluhur Dam in 1970, whereafter sediment yield to the coast diminished sharply. Erosion of the shores of the Po delta has been attributed partly to dam construction and partly to the extraction of sandy material from the river channel, both of which have reduced the sediment yield to the river mouth (Sestini, Jeftic and Milliman 1990).

Erosion on the shores of the abandoned Hwang Ho delta was at first gradual, but it became more rapid about a century after the river diverted into Bo Hai Bay. The delay was due to the gradual conversion of a convex nearshore profile of progradation on the floor of Jiaozhou Bay to a concave profile of erosion.

ω

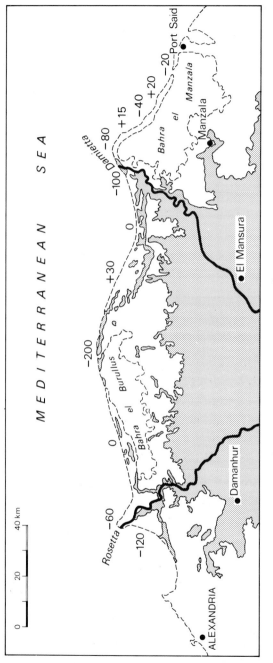

Figure 36 Annual rates of erosion on the shores of the Nile delta since the completion of the Aswan High Dam in 1964 cut off the supply of sediment from the River Nile. Land areas more than 1 m above sea level are shaded, but a sea level rise of this magnitude would cause erosion, and produce a new coastline to landward of this contour. Based on information supplied by G. Sestini.

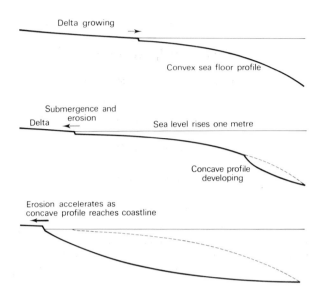

Figure 37   A growing delta typically has a convex nearshore sea floor profile. When sea level rises the delta coast will be submerged and eroded (unless there is sufficient fluvial sediment to maintain it), and increasing wave energy will begin to develop a concave sea floor profile. In due course, when the concave profile intersects the delta coastline, erosion will accelerate

Erosion of a few metres per year accelerated to 200 m per year when the retreating concave profile intersected the delta coastline (Fig. 37).

A rising sea level is likely to cause extensive submergence of low-lying deltaic areas, especially where there is little or no prospect of compensating sediment accretion. A sea level rise of 1 m in such circumstances will result in submergence of land to the 1-m contour, and accompanying erosion will drive the coastline back beyond that line. On the Nile delta, submergence of 1 m would cause coastline retreat of several kilometres (Fig. 36). It has been estimated that a sea level rise of 1 m would inundate much of the deltaic coastal plain of Bangladesh, submerging more than 10% of the country (Milliman, Broadus and Gable 1989); see Fig. 83, page 132.

As sea level rises, progradation of most deltaic coastlines will be curbed, erosion becoming more extensive and more rapid. Sectors of deltas that are already retreating because of submergence and erosion will show accelerated retreat, but progradation will continue around the mouths of rivers that continue to supply sufficient sediment to the coast, including those where the sediment yield has been increased as a result of human activities such as deforestation, farming and mining.

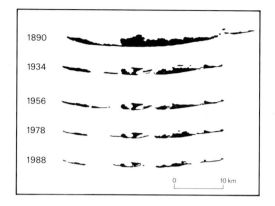

Figure 38   The shrinkage and dissection of Iles Dernières, on the shores of a subsiding sub-delta of the Mississippi (cf. Fig. 36), shows how a rising sea level can modify sandy barrier formations. Based on maps produced by the Louisiana Geological Survey and reproduced by permission of J. Titus, U.S. Environment Protection Agency

Most large deltas, such as that of the Mississippi, already show isostatic subsidence under the weight of accumulating sediment, and submergence of areas no longer being maintained by sedimentation. The Holocene stratigraphy of the Mississippi delta indicates that successive sub-delta lobes built at earlier stages are now partly or wholly submerged and are being dissected by marine erosion (Fig. 34). Margins of sinking delta lobes are indicated by sandy barrier formations which shrink as submergence proceeds (Fig. 38). Coastal subsidence has curtailed the seaward growth of most large deltas, notably those of the Rhine, Guadalquivir, Colorado and Amazon. A rising sea level will increase the effects of subsidence, extending submergence and intensifying erosion.

The extent to which these effects will be countered if sea level rise is accompanied by increasing sediment yields from rivers depends on accompanying environmental changes within the river catchments. Deposition of delta sediments will accelerate if rainfall and runoff from the river catchments increase as a consequence of climatic changes associated with global warming. It has been estimated, for example, that the sediment yield from Javanese rivers will increase by up to 43% as a result of climatic changes leading to greater runoff from their catchments during the coming century (Parry, Magalhaes and Nguyen Huu Ninh 1991). Under these circumstances, sedimentation may build deltas upward to maintain their area as sea level rises, and increased erosion of their coastlines may be offset by the augmented accretion, especially at and around river mouths. It is likely that sedimentation will aggrade natural levees alongside river channels and extend them seaward. Deltas that are now lobate or arcuate in form may become digitate, like the modern Mississippi delta, with intervening areas of submerging swamp and deepening sea.

Much of what has been said about deltas applies to coastal plains that are essentially confluent deltas, as in northern Java. Attention has already been given to coastal lowlands that are beach-ridge plains, but another kind of coastal plain has been produced by emergence of the sea floor, due to land uplift, sea level lowering, or some combination of land and sea movement. The lowlands bordering the Gulf of Bothnia, in Scandinavia, for example, are largely the outcome of emergence due to isostatic land uplift (Fig. 6, page 18). In such areas a sea level rise will reduce the effects of land emergence, slowing down the advance of their coastlines. An accelerating sea level rise will eventually equal, then exceed, the rate of land uplift, and the coastlines will then retreat through submergence and erosion.

INTERTIDAL AND NEARSHORE AREAS

The intertidal zone, defined as the area between high and low tide lines, is wider at spring-tides than at neaps, and reaches its maximum width during the lowest tides. In general, the intertidal zone becomes wider as tide range increases around the world's coastline, attaining 10 km, for example, on the north-west coast of Australia, where Broome has a mean tide range of 8.5 m. The width is also related to the transverse gradient: on the steep rocky coasts of the nearby Kimberley coast, where mean tide ranges exceed 10 m, there are plunging cliffs where the intertidal zone is less than 1 m wide.

The area exposed as the tide falls generally consists of sandy or muddy terrain, with or without rock outcrops, sloping gradually or irregularly down to lowest tide level, in front of cliffs, bluffs, beaches or depositional plains. The nearshore zone is often defined as the area out to the line where waves break at lowest tides, but a more convenient definition is the area where the sea is less than 10 m deep when tides are at a minimum level.

On coasts with low to moderate wave action the upper part of the intertidal zone is usually occupied by vegetation, particularly where the tide range is large and the intertidal zone broad, and where there has been an abundant supply of sediment, especially silt and clay. Salt marshes attain their best development in temperate climates: towards the tropics they may acquire a seaward fringe of mangroves, and within the tropics (especially the humid tropics) they may be entirely replaced by mangroves. Salt marshes and mangroves, discussed in the following sections, have contributed to the shaping of upper intertidal landforms on the shores of inlets and embayments, estuaries and lagoons, and dissipate the energy of waves, thereby protecting their hinterlands, but they fade out as wave energy increases along the coastline.

Salt marshes

Within salt marshes there are often several species arranged in community zones parallel to the coastline, each zone being related to the depth and duration of normal tidal submergence (Ranwell 1972). Landward, they give place to freshwater vegetation such as reeds and rushes, swamp scrub and swamp forest, in a zonation that results from plant succession as accreting sediment builds the substrate up to, and above, high tide level. Vegetation can aid this process, producing patterns of accretion that are more stable than those found in unvegetated areas. Above this level, higher ground is marked by a transition to terrigenous vegetation such as scrub or woodland, or to farmed and developed terrain.

Salt marsh vegetation has spread on to intertidal mudflats and trapped some of the sediment washed to and fro by waves and currents in such a way as to build up depositional terraces on the shores of estuaries, embayments, inlets and lagoons during the Holocene stillstand. These are still prograding where sediment is being supplied at a sufficient rate for their continued upward and seaward development on shores sheltered from strong wave action. Typically, the seaward margin fluctuates: there are sectors of advancing salt marsh and microcliffs where it is cut back (Guilcher 1981). The salt marsh is intersected by tidal creeks into which the sea subsides during the ebb, and drains seaward. When the tide rises, waves and currents carry in silt and clay, together with plant residues and the remains of other organisms such as shells, and when the tide ebbs much of this material is filtered out and retained by the salt marsh vegetation as accreting sediment and incorporated peat. As a result, a salt marsh terrace is formed.

Evidence of such progradation can be found at the seaward edge of some salt marshes. In the Fal estuary, south-west England, a salt marsh advanced seaward by up to 900 m between 1878 and 1973, while its landward edge was colonised by scrub and woodland vegetation (Bird 1985). This occurred during a phase when the river sediment supply was augmented by silt and clay waste derived from kaolin sluicing on Hensbarrow Downs, in the upper catchment.

Changes took place on salt marshes in Britain after a new species, now known as *Spartina anglica*, appeared in Southampton Water in about 1870 as a result of hybridization between local *Spartina maritima* and an accidentally introduced American species, *Spartina alterniflora*. The new species invaded, or was introduced to, salt marshes in other parts of Britain and Europe, and then in many parts of the world. In temperate regions it proved aggressive, displacing pre-existing native salt marsh vegetation and promoting extensive mud accretion so that depositional terraces were aggraded and prograded rapidly. In some areas there has been subsequent 'die-back', and erosion of salt marsh terraces that had been built up under the agency of *Spartina anglica*, which was thus shown to be a plant of geomorphological significance (Ranwell 1964).

On coasts where salt marshes are receiving little, if any, sediment supply, a sea level rise will impede progradation, and increasing wave attack will initiate or accelerate erosion along their seaward margins (Fig. 39). The salt marsh terrace will thus be cut back, and a retreating microcliff will form (Phillips 1986). Tidal creeks which intersect the salt marsh will be widened, deepened and extended headward as the marsh is submerged. As marine submergence proceeds, the retreat of the seaward margin will be matched by a diachronous transgression of the landward margins of the salt marsh on to the hinterland at a rate related to the transverse gradient. Where the hinterland is low-lying, salt marsh vegetation will move landward to displace freshwater or terrigenous vegetation communities. The zonations mentioned previously will thus move landward as they maintain their position in relation to the shifting intertidal zone. Some may widen, others become narrower or coalescent, in relation to variations in the transverse profile of the depositional terrace (Titus 1988).

Figure 39   Erosion and dissection of the seaward margin of a salt marsh in the Bay of Saint Michel, France. It is expected that erosion of this kind will become extensive in salt marshes as sea level rises, except where there is sufficient sedimentation to maintain the seaward margin

Figure 40   An earth embankment built at the rear of a salt marsh near Bosham, southern England, to prevent marine submergence of farmland (right). As sea level rises, such an embankment will prevent the landward migration of salt marsh indicated in Fig. 41

Evidence of such landward migration has been reported from some sectors of the inner margin of salt marshes on the subsiding Gulf and Atlantic coasts of the United States. There are places where hinterland forest is dying as salt marshes encroach in Maryland (Darmody and Foss 1979) and behind the Sounds of North Carolina (Stevenson, Ward and Kearney 1986). Around the submerging shores of Chesapeake Bay salt marsh plants have been invading meadowland. Such landward migration of salt marshes is impeded where the hinterland rises steeply, so that the salt marsh zones are compressed as sea level rises. The same effect will obtain where embankments have been built at the inner margins of a salt marsh (Fig. 40) to keep the sea out of land reclaimed for agricultural and other uses (Boorman, Goss-Custard and McGrorty 1989). Eventually, where sediment accretion is meagre, the salt marsh terrace will cease to exist, and the steep coast or sea wall will be bordered at low tide by mudflats or sandflats (Fig. 41): with a continuing sea level rise, these, too, could become permanently submerged (Park, Armentano and Cloonan 1986).

On the other hand, sedimentation may continue (or be artificially supplied) at a sufficient rate for the depositional terrace to be maintained by vertical accretion as sea level rises (Redfield 1972). The seaward margin will then remain in place,

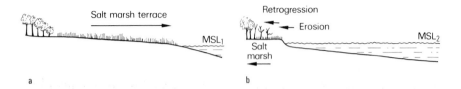

Figure 41   Salt marsh terraces that developed as a result of progradation during Holocene times are likely to be cut back by wave action as sea level rises. The salt marsh is likely to spread landward, replacing other vegetation, if the hinterland is low-lying and not protected by sea walls. (a) Present situation. (b) Response to sea level rise

and existing vegetation patterns could persist (Pethick 1981). As sea level rises there will still be landward migration of the inner salt marsh communities into the hinterland, providing there are suitable low-lying habitats. The outcome will be a widening of the aggrading salt marsh, with variations in species zonation and distribution related to the transverse profile (Reed 1990). Such accretion will be more rapid where the vegetation grows vigorously, especially if such plants as *Spartina anglica* are present, trapping a higher proportion of available sediment and contributing larger quantities of organic (peaty) residues. Coasts with wide bordering mudflats, such as the Baie de Saint-Michel in France and Bridgwater Bay in western England, may well maintain their salt marshes as a result of shoreward drifting of muddy sediment as sea level rises, but the extent to which this will occur is still a matter for debate.

Reference has been made to the salt marshes that form part of the barene or intertidal flats in the Venice Lagoon. These are being rapidly dissected and diminished by erosion (Fig. 42), resulting partly from a rising lagoon level in response to local subsidence (Pirazzoli 1980), and partly from changes in lagoon configuration and hydrology due to bordering land reclamation and fishpond (valli) enclosure during the past century. Other factors have been reduced sedimentation after the diversion of the Brenta River, and increased tidal scour resulting from the dredging and deepening of the Pass of Malamocco.

In southern and eastern England the seaward margins of most salt marshes are generally cliffed and retreating, possibly because of land subsidence along these coasts. Landward migration of salt marshes will almost invariably be prevented by rising hinterlands or embankments built to prevent tidal flooding of reclaimed farmland, so that the area of salt marsh will diminish.

Continuing sedimentation is maintaining the modern delta of the Mississippi River, but the dispersal of sediment over the subsiding deltaic plain has diminished because flood control levees now confine the river and channel its outflow directly into the Gulf of Mexico (Gagliano, Meyer-Arendt and Wicker 1981). It has been calculated that coastal subsidence has been reducing the area

Figure 42   Residual areas of eroding salt marsh in the Venice Lagoon

of Louisiana salt marshes and reedswamps at a rate of 100 km^2 per year: holes develop and grow into rounded ponds that are enlarged by increasing wave erosion as the land sinks. Deterioration of marshlands has been accelerated by the dredging of access canals and extensive disruption in the course of oil exploration and extraction (Salinas, De Laune and Patrick 1986). Salt water intrusion along canals has led to replacement of freshwater cypress forest by salt marshes, but the reduction of floodwater sediment supply behind the levees built to confine rivers has diminished vertical accretion, so that the marshes are being drowned (Day and Templet 1990). In the Savannah River estuary subsidence has resulted in landward migration of salt marshes, which have maintained their area at the expense of freshwater communities displaced upstream (Mehta and Cushman 1989). Sea level rise will accelerate the loss of salt marsh here and elsewhere along the Gulf and Atlantic coasts of North America, where it has been estimated that up to 80% of the coastal wetlands will be inundated if the sea rises by 1 m (Titus 1986b). In New Jersey, Psuty (1992) correlated salt marsh expansion with a slackening of the rate of sea level rise 2500 years ago, and attenuation of marshlands during a phase of accelerated sea level rise within the past century. In Chesapeake Bay some marshy islands have diminished in area and others have disappeared during this latest phase (Kearney and Stevenson 1991).

The stratigraphy of sediments beneath some salt marshes indicates that they were maintained with the aid of vertical accretion during the Late Quaternary marine transgression and that they expanded seaward when that transgression came to an end. There is often stratigraphic evidence of the seaward advance of successional zones (Ranwell 1972). On coasts that are still subsiding, as in New England, salt marshes exist where they have aggraded at a sufficient rate to match the relative rise of sea level (Redfield 1972). Where the aggradation is slower than the sea level rise, as in parts of New Jersey and Louisiana, salt marshes are being drowned and eroded, but where it is more rapid than the sea level rise they are developing and spreading (Stevenson, Ward and Kearney 1986). Near Philadelphia, Orson, Panagetou and Leatherman (1985) found vertical accretion in salt marshes at the rate of 1.0–1.2 cm per year during the past 50 years on a site where sea level had risen only 0.265 cm per year over that period, implying that aggradation was still 'catching up' with submergence in this region. Sudden subsidence along the shores of Turnagain Inlet, Alaska, during the 1964 earthquake resulted in replacement of spruce forest by salt marshes (Fig. 43).

Figure 43   A former spruce forest has been killed and replaced by salt marsh on the shores of Turnagain Inlet, Alaska, as a result of sea level rise due to subsidence during the 1964 earthquake

On the coasts of the Baltic Sea, where salinity is low and there is little tidal movement, freshwater reeds and rushes form communities on sheltered sectors of the shore and are spreading into shallow nearshore waters. These also trap sediment and incorporate peaty material in swamp land that progrades in a similar way to salt marshes, building up a depositional terrace to a level where it is colonised by terrigenous scrub and forest vegetation. In the Gulf of Bothnia this reed and rush encroachment has been aided by coastal emergence due to isostatic land uplift. A rising sea level will slow this down, and eventually reverse it, leading to erosion, submergence, and landward migration of these coastal wetlands (Bird 1989b).

Mangroves

Mangroves grow in tropical and subtropical environments, and extend locally into the temperate zone: in the southern hemisphere they occur as far south as the southern shores of Corner Inlet, in Australia, at latitude 38°49′. In the humid tropics they grow luxuriantly, especially where rivers have been delivering large quantities of silt and clay to prograde the coast. They form tree communities in the upper intertidal zone, backed by tropical rain forest or land cleared for agriculture or other development. They attain their greatest diversity in parts of south-east Asia, notably in Indonesia, where there are more than 30 species, and are often zoned in distinct ecological communities, forming forests up to 40 m high intersected by branching tidal channels. They are very extensive on the north-east coast of Sumatra and the southern coast of Irian Jaya. In the Caribbean and Latin America there are fewer species, and simpler communities.

On subtropical and temperate coasts mangroves generally narrow to a lower, scrubby seaward fringe, typically backed by salt marshes, as in Westernport Bay, Australia (Bird and Barson 1975), with zones of freshwater vegetation and swamp scrub or reclaimed land to the rear. On arid or semi-arid coasts the mangrove fringe is usually backed by unvegetated salt pans, traversed by mangrove-fringed tidal creeks, with savanna or desert hinterlands.

Mangroves grow down to about mean tide level. Where the sea is shallow and the coast sheltered, they spread directly on to accreting tidal mudflats, but where wave action is stronger the seaward fringe often consists of a narrow protective beach of sand. This is the case, for example, along much of the west coast of Peninsular Malaysia, and in Sarawak and Sabah.

Like salt marshes, mangroves trap sediment to build a depositional terrace in the upper intertidal zone, and as they do so, successive zones of one or more species migrate seaward (Fig. 44). As the terrace is built up to high tide level, other vegetation, such as rain forest, moves in to displace the inner mangroves. This evolution depends on a continuing sediment supply, as on deltaic shores close to the mouths of rivers. On shores now receiving little or no sediment,

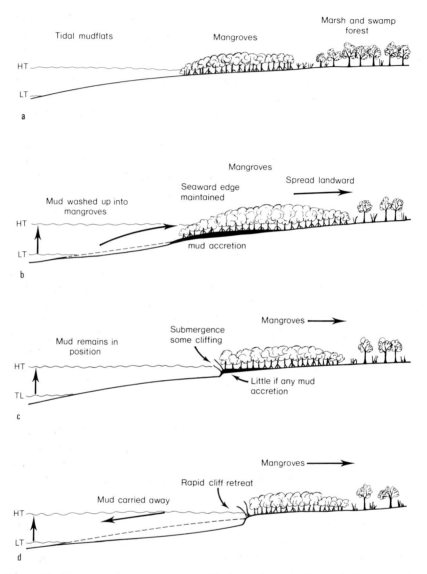

Figure 44   Mangroves have spread seaward in front of salt marshes and other vegetation on terraces formed by sediment accretion on the tidal shores of bays, estuaries and lagoons. (a) Present situation. (b)–(d) As sea level rises there are three possibilities: (b) that there will be sufficient mud accretion to maintain the seaward edge while the mangroves spread landward; (c) that in the absence of a sufficient sediment supply the seaward margin will become cliffed as the mangroves spread landward; (d) that removal of sediment and nearshore deepening will lead to stronger wave action and rapid retreat of the seaward cliff as the mangroves spread landward. HT, high tide; LT, low tide

mangrove terraces built by the re-shaping of intertidal topography have become relatively stable, or are being eroded. Low receding cliffs have been cut into the seaward margin of mangrove terraces, particularly on deltaic coasts where the sediment supply has been reduced because of dam construction upriver or the natural or artificial diversion of a river outlet. Mangrove terraces are also trimmed back by meandering estuary channels, or where the nearshore zone has deepened as a result of shoal migration, and larger waves are moving in to the mangroves.

Mangrove communities have been extensively modified by human activities, especially during the last few decades. They have been destroyed in the course of land reclamation for agriculture, urban and industrial development and port construction. In south-east Asia there has been large-scale reclamation of mangrove areas for land development around such places as Penang and Melaka in Malaysia and Jakarta and Surabaya in Indonesia. Elsewhere, mangroves have been depleted by cutting for timber and fuel wood, especially in Arabia and Africa, or cleared in order to dredge out placer deposits, such as tin, which occur beneath mangroves on the west coasts of Thailand and Malaysia. The most extensive clearance in recent decades has been made for the establishment of aquaculture (fish and prawn ponds) and salt pans in such countries as Ecuador and Brazil, and especially in south-east Asia, where destruction of mangroves has been very extensive. In Thailand, for example, reclamation and aquaculture have reduced the mangrove area to less than 40% of its natural extent (Aksornkoae 1988), and similar reductions have occurred in Malaysia and Indonesia. Globally, mangrove losses have already been substantial, even before a global sea level rise becomes effective.

As in salt marshes, the effects of a sea level rise on remaining mangrove areas will depend on the extent to which they continue to receive sediment. Where there is little or no sediment supply, submergence is likely to cause die-back and erosion of their seaward margins, and the mangrove zones will migrate landward if low-lying hinterland sites are available (Fig. 44). A sea level rise will thus reverse the successional advance of mangroves that took place as accretion was shaped into depositional terraces on prograding coasts. It is likely that some species will become more abundant and extensive, while others diminish and even disappear. As their seaward margin is cut back, mangroves will spread landward to displace freshwater swamp or forest on low-lying estuarine and alluvial terrain, but in many areas the inner margin has been defined by a bank built to protect reclaimed land, and if this is maintained the mangrove fringe will become narrower, and eventually disappear as sea level rises (Fig. 45). The area of mangroves will also diminish in front of banks that enclose ponds built for the farming of fish or prawns, or ricefields (Armentano, Park and Cloonan 1986, Titus 1986b), and as the mangroves disappear, these banks will be exposed to wave attack. They are likely to be armoured to prevent marine incursion, so the coastline will become more and more artificial.

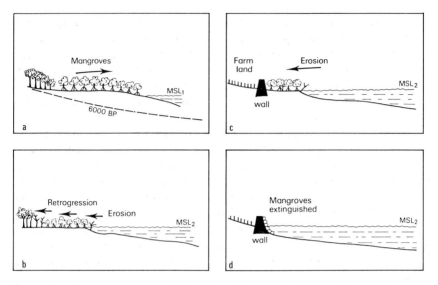

Figure 45   Changes on mangrove-fringed coasts as sea level rises (a) and (b) will be modified where they are backed by a wall built to protect farmland (c). The mangrove fringe will then be narrowed by erosion, and may eventually disappear (d), unless there is a sufficient sediment supply to maintain the substrate and enable the mangroves to persist

As with salt marshes, mangroves may persist, or continue to prograde, where the rate of substrate accretion equals or exceeds the rate of sea level rise, particularly in the vicinity of the mouths of rivers supplying sediment (Woodroffe 1990). There is also a possibility that as sea level rises, some of the sediment stirred from bordering mudflats by strengthening wave action will be washed up into the mangrove fringe. This would result in aggradation of the depositional terrace beneath the mangroves and widening of the mangrove fringe where its inner margin can transgress landwards. Where intertidal areas are enriched with nutrients derived from eroding sediments or enhanced runoff from the hinterland it is possible that invigorated growth of mangroves could match a sea level rise by building up the depositional terrace with accumulating mangrove peat. In the absence of sustaining sediment accretion, it is thought that mangroves could maintain themselves on their accumulating peat with sea level rising at up to 9 cm per century, but that they would be impeded by a faster submergence, and collapse when the rise exceeded 12 cm per century (Ellison and Stoddart 1991). In the Gulf of Papua, Pernetta and Osborne (1988) decided that fluvial deposition on the Purari delta would provide sufficient sediment to maintain the mangrove ecosystem as sea level rose, whereas in the tidally dominated estuaries of the Kikori coast, where sedimentation is slower, they would retreat landward and become reduced in area.

Mangroves have become established on some low coral reef islands, such as the Low Isles off north Queensland, where they occupy shallow basins with thin veneers of muddy sediment. They are unlikely to survive as sea level rise proceeds because rates of sediment accretion are very low in such areas, remote from rivers and terrigenous sources (Ellison and Stoddart 1991). The same is true of mangrove islands with a peat and mud foundation in the north-west of Torres Strait (Mulrennan 1992).

The response of mangroves to sea level rise can be examined on subsiding coasts. At Port Adelaide in South Australia, where local tide gauge records indicate that a relative rise of sea level has been taking place, mangroves spread back into salt marshes at a rate of 17 m per year between 1935 and 1979 (Burton 1982). Similar changes have been found on the low-lying marsh islands of Corner Inlet, an area of tectonic subsidence in south-eastern Australia, where the mangroves are displacing submerging salt marshes (Fig. 46) (Vanderzee 1988), and on the Mamberano delta, in Irian Jaya, where fringing mangroves have spread back into a subsiding corridor (Fig. 47) (Bird and Ongkosongo 1980).

On the coast of West Johore, in Malaysia, the seaward margin of the mangroves has been cut back by erosion, so that the depositional terrace ends

Figure 46 Mangroves spreading back into salt marsh on low-lying muddy islands in Corner Inlet, south-eastern Australia, where sea level is thought to be rising because of tectonic subsidence of the coastal land

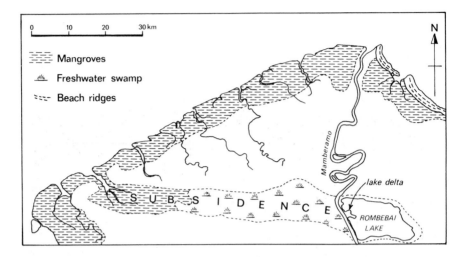

Figure 47   The abiliity of mangroves to retreat as sea level rises is indicated on sectors where subsidence of the land has occurred. On the Mamberamo delta, Irian Jaya, Indonesia, mangroves have spread back into the western part of a subsiding corridor across the southern part of the delta

in a sharp cliff along the seaward side. Such cliffs are formed during occasional storms, and are apt to persist because onshore movement of muddy sediment is impeded where they have formed, the cliff reflecting wave action and promoting scour of the fronting mudflats. Once such a cliff has formed it will retreat, consuming the mangrove fringe, unless there is substantial accretion of muddy sediment from offshore or alongshore to restore the mudflats to the level of the cliff crest. A rising sea level will accelerate this retreat.

On the shores of the Gulf of Thailand mangroves are spreading forward on to accreting mudflats at Bangpu (Fig. 48), in the most rapidly subsiding sector of the Bangkok coast (Fig. 49), but eroding on the coast to the west, which is not subsiding. This is the reverse of the expected response to sea level rise, and the explanation is that mud dredged from the Chao Phraya River to maintain navigation to the port of Bangkok has been dumped off the Bangpu coast, and has been washed onshore by wave action, unintentionally renourishing the mudflats, and more than offsetting the effects of coastal subsidence (Vongvisessomjai 1990). The implication is that mud could be dumped to renourish coastal swamps in much the same way as sand dumping has renourished beaches.

Where evidence from the Holocene stratigraphy beneath mangrove areas is available, it generally indicates that they spread to their present extent during the sea level stillstand of the past 6000 years. Before that, during the Late Quaternary marine transgression, they were confined to sheltered inlets and estuarine sites

Figure 48   Mangroves advancing on to tidal mudflats at Bangpu, south of Bangkok (cf. Fig. 49). Although the land has been subsiding here, mud accretion has maintained the substrate at a level where mangroves can persist, and spread seaward

where they could migrate landwards, and to sectors where they could persist on vertically accreting muddy substrates as submergence proceeded. Stratigraphic studies in northern Australia and south-east Asia have shown that mangroves were growing in accreting estuaries between 7000 and 5500 years ago, and that when the Late Quaternary marine transgression came to an end they spread out to other embayments and more exposed sites of muddy accretion along the outer coast (Fig. 50) (Woodroffe 1988, Woodroffe, Thom and Chappell 1985). A sea level rise will reverse this sequence. Mangroves will disappear from the more exposed coasts, and retreat behind sheltered inlets and embayments, persisting only where accretion is sufficient to permit their continued presence, mainly in estuaries receiving large quantities of fluvial silt and clay. Some mangrove species will prosper, but others may diminish and could be extinguished as a result of sea level rise.

Intertidal mudflats, sandflats and rocky areas

Seaward of salt marshes and mangroves, and on shores where these are absent, mudflats, sandflats and rocky areas exposed at low tides are unvegetated, or carry a variable cover of seagrasses, marine algae, and other plants and associated

86

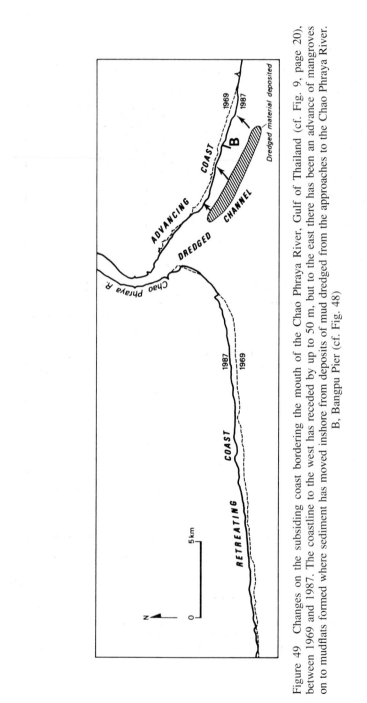

Figure 49 Changes on the subsiding coast bordering the mouth of the Chao Phraya River, Gulf of Thailand (cf. Fig. 9, page 20), between 1969 and 1987. The coastline to the west has receded by up to 50 m, but to the east there has been an advance of mangroves on to mudflats formed where sediment has moved inshore from deposits of mud dredged from the approaches to the Chao Phraya River. B, Bangpu Pier (cf. Fig. 48)

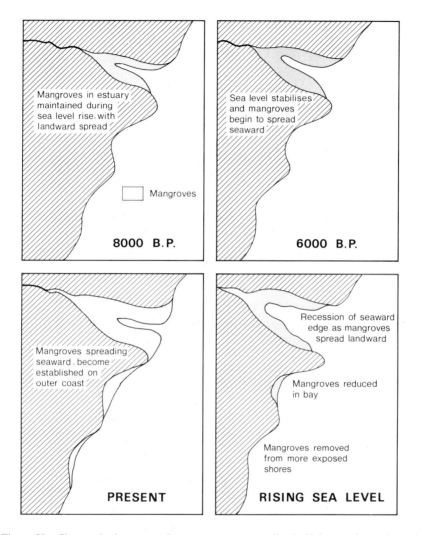

Figure 50   Changes in the extent of mangroves on a coastline in Holocene times. Around 8000 years ago, when sea level was rising, mangroves were generally confined to estuaries where mud accretion kept pace with the rising sea. As the marine transgression ended about 6000 years ago the mangroves began to spread, and now occupy embayments on the outer coastline. If sea level rises, mangroves will be reduced on the outer coastline, but may persist in muddy estuaries where accretion maintains their substrate

animals. In some areas there are shelly organisms, such as cockles, mussels and oysters, and shelly debris derived from these can form gravelly shoals that may be washed on to backing beaches. Sandy and muddy sediments are moved to and fro by waves and currents, and arranged into bars, shoals and ripples of variable configuration, usually traversed by tidal creeks and seepage corridors of softer, wetter sediment sloping down to low tide level.

These areas will be modified as sea level rises. The outer part of the intertidal zone will become permanently submerged, and as salt marshes and mangrove swamps are eroded, and coastal lowland fringes cut back, the intertidal zone will extend landward. Areas previously occupied by salt marsh or mangroves will become mudflats or sandflats, and underlying rocky areas may be exposed. Landward migration of this kind will not be possible where the coast rises steeply, where sea walls have been built, or where parts of the intertidal zone have been enclosed by sea walls for land reclamation. In these areas the intertidal zone will be reduced in width, and eventually submerged by the rising sea.

Seagrasses occupy parts of the intertidal and nearshore zones. They require a suitable substrate, generally mud or fine sand, and light penetration limits them to low tide depths of less than 10 m. Their growth is impeded by strong turbulence, and where they have been destroyed the release of substrate sediment may render the water turbid, and less favourable for the growth of corals and other marine organisms. Seagrass growth is improved by accessions of fine-grained sediment and the increase in nutrients that results from non-toxic pollutants, but if nutrients become excessive, eutrophication can cause blooms of algae, which can blanket and destroy seagrasses. Growth of seagrasses also promotes accretion of sediment and organic materials, and can diminish wave action, thereby helping to stabilise the adjacent coast, but where they have been destroyed wave action is less impeded, and wave attack intensifies on the coastline. Marine algae have generally similar effects to seagrasses, but can grow on coarser sediment and rocky outcrops, and extend into water tens of metres deep.

As sea level rises seagrasses and marine algae will tend to migrate landwards, their inner limits spreading on to the submerging sandy and muddy substrates that form as beaches are submerged and salt marshes and mangrove swamps are eroded, their outer margins dying away as the water becomes too deep. Increased wave energy as sea level rises may generate greater turbidity, thus reducing sunlight penetration and impeding plant growth on the sea floor. On the other hand, it is possible that seagrasses and other intertidal and nearshore plant and animal communities will become wider and more luxuriant on coasts where broad plains, rich in soil nutrients, are being submerged by the rising sea.

There has been extensive embanking, draining and reclamation of intertidal zones, including salt marshes and mangroves, around the world's coastline, mainly for agriculture or urban and industrial development, including ports.

Around the Mediterranean, for example, there are now only a few areas where coastal marshes and tidal mudflats or sandflats persist (Sestini, Jeftic and Milliman 1990).

As sea level rises, intertidal and nearshore areas will migrate landwards, their widths being modified in relation to the transverse gradient, which in turn may steepen in response to strengthening wave action as the sea deepens. In general, changes in the features of intertidal and nearshore zones as sea level rises will depend on the extent to which submergence is offset by continuing sediment accretion, and on the extent to which these zones can migrate landward. The sediment supply is more likely to be maintained in the vicinity of river mouths and may increase where fluvial sediment yields are augmented by larger or more intensive rainfall and runoff; it could also be maintained where there are shallow sea areas from which sediment can drift shoreward as submergence proceeds. Landward migration can occur particularly where the coast is low-lying, and not protected by sea walls. On the world scale, however, it seems likely that a rising sea level will greatly reduce the extent of intertidal zones, except where they can move back into low-lying hinterlands, which may have to be made available for this purpose (Pernetta and Elder 1990). Nearshore zones, on the other hand, are likely to persist during submergence, and could widen off many coasts as transverse profiles are re-shaped. However, the response to a rising sea level could also include more extensive coastal land reclamation, invading and reducing intertidal and even nearshore areas.

Coral reefs

Corals and associated organisms, notably calcareous algae, have built extensive reefs in sea areas within a zone extending 30° north and south of the equator, especially in the western parts of the Pacific, Indian and Atlantic Oceans. They include barrier reefs that run parallel to mainland coastlines, atolls encircling lagoons, miscellaneous patch reefs, and fringing reefs on the shores of headlands and high islands.

Each of these reefs has been built by coral polyps, small marine animals that extract calcium carbonate from sea water and grow into a variety of skeletal structures to form a framework within which the growth of calcareous algae and other organisms, and the deposition and precipitation of sediments, produce a solid rock formation. Corals can flourish, and initiate such structures in clear warm seas, where the temperature of the coldest month does not fall below 18°C, and salinity remains within the range of 27–38 parts per thousand. Their growth is impeded where the temperature rises above about 28°C, where the sea becomes hypersaline, as in parts of Shark Bay, Western Australia, or where it is diluted by fresh water off river mouths. As coral growth is also inhibited by turbidity and sedimentation, reefs are usually poorly developed, or absent, off river mouths and along deltaic shores where there is much suspended terrigenous silt and clay in the sea.

Active coral growth occurs to depths of at least 50 m in clear, warm water, but coral reef formations extend to much greater depths than this. Darwin (1842) proposed his subsidence theory, whereby existing reefs had been initiated as fringing reefs along coastlines and around islands, and that as these foundations subsided tectonically the fringing reefs had grown up to persist as barrier reefs and atolls with configurations that commemorate ancient coastlines. Subsequent research has refined this picture, acknowledging that there have been eustatic oscillations of sea level as well as tectonic movements (Daly 1934, Dana 1872, Davis 1928).

Coral reefs have thus attained their existing form as the result of upward growth during and since the Late Quaternary world-wide marine transgression, but most coral reefs have had a much longer history. Many have developed during the successive marine regressions and transgressions that accompanied the waxing and waning of Pleistocene glaciers and ice sheets, and before that the longer-term oscillations of land and sea level that have occurred through geological time.

The northern part of the Great Barrier Reef in Australia, for example, has foundations which embody early Tertiary reefs, and is a structure which has been repeatedly exposed to atmospheric weathering and erosion during low sea level phases, when it became an intermittent ridge of dissected limestone at the outer edge of the coastal plain which is now submerged as the Queensland continental shelf. During intervening transgressions corals recolonised the karstic reef limestone ridges, and grew to enclose and cap them with younger reef material. This culminated during the Late Quaternary marine transgression with further reef upgrowth from eroded foundations now 10–15 m below sea level, and during the Holocene phase of relative stillstand the reefs have widened to their existing outlines (Hopley 1982). Most oceanic reefs have had similar lengthy histories, but fringing reefs are generally younger and simpler, and some may be entirely of Late Quaternary origin.

As corals cannot endure prolonged exposure to the atmosphere, most coral reefs have been built up to only just above low tide level, where they consist of solid reef flats exposed at low tide (Fig. 51), with mainly dead corals and various algae (such as *Porolithon* spp.), bordered by living and growing coral ecosystems in the intertidal and subtidal zones. Living corals on reef flats are few and scattered, and if they are killed they are not quickly replaced.

The solidity of coral reef flats in the Pacific and Indian Oceans implies not only the attainment of present intertidal levels by upwardly growing corals, accompanied by algal growth and sedimentation, but also a phase of secondary cementation (Guilcher 1988). On the Great Barrier Reef, Hopley (1982) found that radiocarbon dates of the upper few centimetres of coral reefs were typically

Figure 51   Reef flats exposed at low tide on the Great Barrier Reef, Australia, consist largely of dead coral and algae, the live corals being on the submerged reef flanks

within the past 2000 years, indicating that there had been very slow upward growth. There may even have been some truncation, especially in the Pacific region where coral reefs built up to a higher Holocene sea level, 1–2 m above the present, about 6000 years ago (Chappell 1982, Pirazzoli and Montaggioni 1986) have since been eroded down to their present horizons. By contrast, many of the reefs of the Caribbean region, where sea level has continued to rise throughout the Holocene, are more fragile 'reef gardens', which have not been able to attain the solidity of Indian and Pacific Ocean reef flats because of the lack of a higher Holocene sea level and ensuing emergence.

Corals are now growing vigorously on the flanks of many reefs below low tide level, widening reef flats that have attained their upward limits. There is compact coral growth on windward aspects, washed by strong surf, where the nutrient supply is abundant, and extensive areas of upward growing coral on the lee side of barrier reefs and within atoll lagoons. There is much variation in the species composition and distribution of corals, algae and other organisms, and in the extent and manner in which they have been modified by human activities. Existing reefs have also been modified by the impacts of tsunamis, hurricanes and earthquakes (Guilcher 1988).

A rising sea level is expected to lead to a revival of upward reef growth, initiated by the expansion and dispersal of presently sparse and scattered living corals on reef flats. If global sea level has been slowly rising during the past century at 1.0–1.2 mm per year, and present sea level is therefore 10–12 cm higher than it was in the 1890s, it could be expected that coral reefs, regarded as sensitive to such fluctuations, would be among the first features to show indications of such a sea level rise, but there have not yet been reports of the general spread of living corals or of accelerating coral growth (Bird 1988b). Few reefs have been mapped and monitored with sufficient accuracy for such a change to be measured, but a widespread revival of coral growth on reef flats should have been noticed. Perhaps a threshold has yet to be reached in terms of the rate of sea level rise before such a response will occur (Buddemeier and Hopley 1989); perhaps coral revival could be stimulated artificially by planting live corals on reef flats.

Revival of coral growth with a rising sea level will be strongly influenced by ecological factors, which influence the ability of coral species to recolonise the submerging reef flats, and could be delayed or inhibited where these factors are adverse. There is no doubt that coral reef ecosystems have been very widely modified by human activities, especially during recent decades. The most obvious damage has been caused by the use of explosives to harvest fish, and by fishermen beating the shallows to drive fish into nets. The use of chemicals and gases to capture fish has also had adverse ecological consequences. Other reefs have been quarried for sand, gravel and building stone, or damaged by the cutting of boat access channels, or by boat anchors. Such activities generate sediment turbidity, which also impedes coral growth in neighbouring areas.

Many reefs have been affected by pollution, including the in-washing of muddy sediment generated by nearby dredging, or material dispersed during drilling for oil, or oil spills from wrecked or damaged tankers. In recent decades, deforestation of hinterlands has increased sediment flow into coastal waters, raising turbidity and killing many fringing and nearshore coral reefs by blanketing them with sediment. Around Sulawesi, in Indonesia, coral reefs that were rich in species have been impoverished by sedimentation resulting from soil erosion, the increased turbidity having made them less vigorous, and reduced the number of species, especially near coastal towns. Excessive nutrients from eroding soils, agricultural fertilisers and sewage pollution have caused eutrophication, which is detrimental for coral growth: for example, there is a correlation between high nitrogen and phosphorus levels and reduced calcification of coral reefs (Kinsey and Davies 1979). Eutrophication also results in the killing of corals by the growth of algae and other organisms. Collecting of shells and precious corals has also impoverished many reefs, and in some countries Marine National Parks have been set up in an attempt to protect coral ecosystems from these impacts, but the products are still widely on sale, and such controls are not yet effective. Outbreaks of destructive crown-of-thorns starfish on coral reefs may be a response to human activities, but some have interpreted them as a natural cyclic phenomenon.

Global warming will modify the distribution of corals by increasing sea temperature and salinity in areas that become more arid. Recent reports of widespread coral bleaching (Brown 1990, Williams and Williams 1990) may be a consequence of higher temperatures in tropical seas, notably during the ENSO in 1982–1983 and 1987, when coral mortality was extensive in the eastern Pacific. They are a reminder that increasing sea temperatures are likely to impede coral growth, especially certain species such as the staghorn corals (*Acropora* spp.), and also modify the productivity of algae such as the sand-producing *Halimeda*, which plays an important role in reef sedimentation.

Climatic changes that result in higher rainfall and greater discharges of fresh water will also diminish coral growth, especially where runoff from the hinterland deposits increasing amounts of sediment on coral reefs.

The response of coral reefs will depend on the rate at which sea level rises. A slowly rising sea should stimulate the revival of coral growth on reef flats, but an accelerating sea level rise will lead to the drowning and death of some corals, and eventually the submergence of inert reef formations. Neumann and Macintyre (1985) pictured the relationship between rates of sea level rise and rates of upward reef growth in terms of keep-up, catch-up, and give-up reefs, noting that there will be variations in response related to existing morphology and depth, as well as ecological composition. As sea level rises there will be an accompanying rise in the depth limit for coral growth, so that the deepest corals die. Theoretically, corals could also expand northward and southward beyond their present latitudinal limits as sea temperatures rise, but this is likely to be a slow response because of impeding factors such as those resulting from human impacts (Hopley 1989).

Several authors have emphasised the difference between the growth rates of individual organisms forming a reef (which can be quite rapid, the branches of some *Acropora* species having extended several centimetres per year) and the reef formation as a whole (Hubbard 1985). Measurements of mean growth rates of existing corals are in the range 0.4–7 mm per year (Hopley and Kinsey 1988), with up to 10 mm per year in favourable conditions (Buddemeier and Smith 1988), and it can be argued that reef formations will grow upward to match sea level rise as long as it is within this range. Similar conclusions on rates of potential reef growth have been reached from analyses of carbonate budgets in coral ecosystems (Scoffin et al. 1980) and chemical indices such as alkalinity depression of reef waters as a consequence of carbonate extraction by corals and associated reef organisms (Smith and Kinsey 1976).

Coral species will respond variously to sea level rise (as well as to increasing water temperature and perhaps storminess), and it remains to be seen which will prosper as the sea rises. Reefs which sustain rapidly growing corals are more likely to be maintained, but as sea level rise accelerates it will drown first the slow-growing coral species, and eventually the whole reef ecosystem. A slackening of sea level rise would permit some surviving corals to grow up, and others to recolonise as the reef shallows. Reefs submerged by up to several metres after a period of rising sea level could catch up if the rate of submergence then diminished, and especially if a new stillstand were to ensue.

The impacts of a sea level rise on existing coral reefs can also be considered in terms of stratigraphic evidence of what happened to coral reefs during the Late Quaternary marine transgression. It should be borne in mind, however, that the ecology and geomorphology of existing reef formations differ in various ways from those when Late Quaternary reef revival commenced on submerging coastlines and pre-existing dissected reef limestones, and a much cooler climate had started to ameliorate. The most notable differences are the impacts of human activities on existing coral reefs during the past few centuries. In view of these impacts it may be that the world's coral reefs are much less capable than they were under the natural conditions of the Late Quaternary to respond to a sea level rise (Yap 1989).

Studies of reef evolution during the Late Quaternary marine transgression, when the sea rose at an average rate of about 1 m per century, indicate typical rates of upward growth in the range of 1–8 mm per year (Davies 1983). This enabled a variety of growing corals in the reef framework to be within a few metres of sea level when the transgression slackened about 6000 years ago, and then to extend upward and outward to form existing reefs. Where coral growth failed to attain a sufficient rate, reef structures were drowned. Davis (1928) mentioned examples of drowned barrier reefs off the south-eastern coast of New Guinea and in the Fiji archipelago, and there are examples of drowned atolls, notably in the Caroline Islands (Guilcher 1988).

Evidence of the response of corals and coral reefs could be obtained from areas where reefs are known to be subsiding, but these have not been well documented. Alternatively, evidence may be sought from studies of tilting coral reefs and atolls, such as Uvea in the Loyalty Islands, where the submerged portion shows rapid, if patchy, upward growth of corals. A model of coral response on tilting reefs shows zones of coral growth responding well to slow submergence (<5 mm per year) where upward reef growth is being maintained, passing laterally to zones of more rapid submergence (5–10 mm per year) where the corals are growing, but are failing to 'keep up' with the rising sea, and to zones (>10 mm per year) where the corals have died and the reefs are drowned and inert (Fig. 52).

The evidence from the submerging portion of Uvea (Fig. 53) is a reminder that upward growth of corals does not immediately form a solid reef flat; instead, there are growing 'reef gardens', rich in corals and accumulating sediment, but fragile structures that cannot be walked over. As existing Pacific and Indian

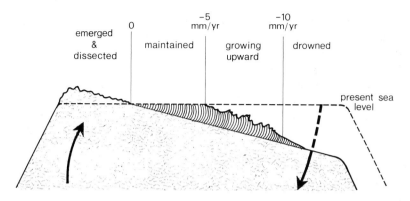

Figure 52   The response of coral reefs to a sea level rise may be predicted with reference to changes on tilting reefs or atolls. In this model, reefs are being maintained by upward growth of corals, associated organisms and sedimentation in the zone subsiding less than 5 mm per year, are still growing upward in deepening water in the zone subsiding at 5–10 mm per year, but are drowned where the subsidence rate exceeds 10 mm per year

Ocean reef flats are submerged by the rising sea, reviving coral growth is likely to form ecosystems similar to those found in the Caribbean, or in shallow water to the lee of existing reefs, rather than consolidated reef flats, which evidently require the attainment of sea level stillstand for their completion, or even a phase of down-cutting into established reef limestone.

Pelsart Reef, on Houtman Abrolhos, off Western Australia, is an example of a fragile 'reef garden' of the kind that will characterise a keeping-up reef ecosystem as sea level rises. Fairbridge (1947) cited this as a possible example of a coral reef responding to a globally rising sea level, but it is an isolated example, which could be the outcome of local subsidence or, more likely, of impeded reef evolution in an area close to the limit of coral growth (Smith 1981).

It has been pointed out that localised 'moating' of water levels behind accumulating shingle ramparts or algal rims can lead to the formation of a slightly higher level of living coral in the form of flat-topped micro-atolls without any evidence of sea level change (Hopley 1982, Scoffin and Stoddart 1979). This implies that a positive response would occur in coral growth if sea level rise became general (Woodroffe and McLean 1990). It is a localised phenomenon that confirms the lack of a general revival of corals on reef flats, mentioned previously.

Experimental transplanting of various corals to deeper levels, as on the artificial reef at Sentosa in Singapore, may elucidate the response of various species to sea level rise. Apart from corals, the ability of reef algae to respond to sea level changes was shown by Cubit (1985) with reference to migrations of

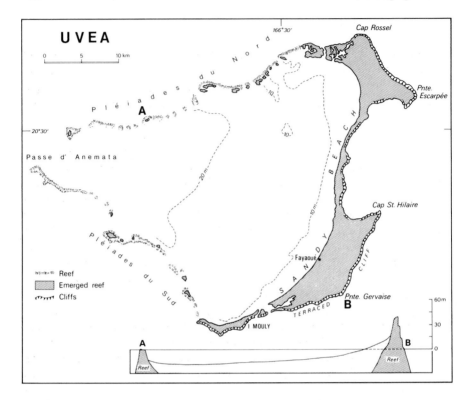

Figure 53   The tilted atoll of Uvea, Loyalty Islands, New Caledonia, shows an emerged reef on the eastern side and intermittent reefs, the Pléiades du Nord and Pléiades du Sud, growing up from the subsided atoll rim to the west. The Pléiades include segments that have grown up to match the relative sea level rise, with intervening sectors of living reef below low tide level, and a drowned reef beneath the western Passe d'Anemata.
Compare Fig. 52

Laurenica papillosa across reefs at Punta Galena, in Panama, correlated with annual fluctuations of local mean sea level.

As has been indicated, many coral reefs are already under various kinds of ecological stress, with diminished species numbers and reduced vigour of growth. Some will fail to revive, and become permanently submerged as sea level rises. Fringing and nearshore reefs are less likely to persist than outlying reefs because of stronger human impacts, increasing turbidity in coastal waters, and more intensive erosion as larger waves attack the coast. Where there is a greater frequency of tropical cyclones such erosion will accelerate, and increased rainfall over hinterlands will augment fluvial sediment yields and nearshore turbidity. A rising sea will submerge fringing reefs in much the same way as rocky shore platforms fronting steep and cliffed coastlines.

Islands on Coral Reefs

Some reef flats are surmounted by small islands (cays) of coralline sand and gravel eroded from the surrounding reef and washed up by wave action to be deposited in locations and configurations determined by wave refraction over the reef. There are also elongated islands (motus) formed where reef limestone blocks and rubble have been thrown up by storms close to reef margins. These features rise only a metre or two above high tide level, and carry land vegetation, usually modified by human settlers. They are well developed on reefs in the Indian and Pacific Oceans, where their evolution has been facilitated by emergence following the attainment of a higher Holocene sea level and the ensuing reef erosion, and where there is often protection by prominent algal ridges along the windward margins. In the Caribbean, where there was no higher Holocene sea level, the more numerous islands are smaller, more scattered, and more vulnerable to erosion (Stoddart 1990, Stoddart and Steers 1977). The tide ranges are small, and the islands rise above 'reef gardens' that are generally submerged and overwashed by wave action, which generates ample supplies of coralline sediment from the reef environment.

Low islands on reef flats already show evidence of changes in configuration, even with a relatively stable sea level. Associated beach rock and conglomerate, formed where carbonate precipitation has cemented coralline sand and gravel in the zone of beach water fluctuations, have impeded erosion, but exposed patterns of relict beach rock and conglomerate indicate former positions of low islands that have migrated, usually to leeward, or vanished altogether. Stoddart (1971) documented the loss of several cays on the Belize barrier reef in the Caribbean during Hurricane Hattie in 1961, noting that diminution of their vegetation by man during the previous century and a half had allowed hurricanes to become more destructive.

Such changes will be accentuated by a rising sea level, when many low islands will be eroded by larger waves approaching through deepening waters, and may disappear, overwashed by storm surges (Fig. 54). There is also likely to be an increased frequency and severity of tropical cyclones as sea and atmospheric temperatures rise (Emanuel 1987). On the other hand, there is a possibility that, where submergence is slow enough, reviving coral growth on the surrounding reef flats will at least partly offset erosion of low island shores by impeding wave attack. The islands may even be enlarged by accretion of coralline material derived by stronger wave action from the growing 'reef garden' ecosystem (Fig. 54).

There are thus difficulties in predicting what will happen to coral reefs and reef islands if sea level rises in the manner expected. While the impacts of sea level rise on coral reefs and reef islands can be outlined in general terms, the influence of ecological factors and the effects of human interference are more difficult to assess. It is surprising that coral reefs do not already show a positive response

PRESENT FEATURES   **AFTER SEA LEVEL RISE**

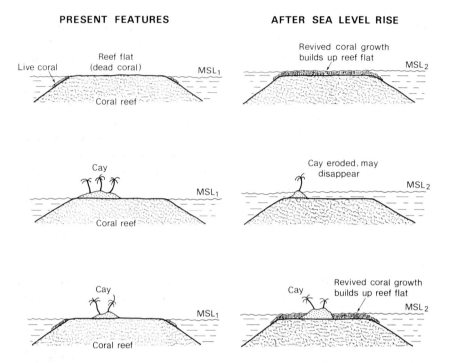

Figure 54 The response of a coral reef to a rising sea level. If coral growth revives, it may build up the reef flat to match the sea level rise (top). If there is an island of coralline sediment (cay) on the reef, it may be eroded and eventually destroyed by the rising sea (middle), but if coral growth revives to build up the surrounding reef flat the cay may be preserved, and even enlarged, as sea level rises (bottom)

to the rising sea level that many authors have deduced from analysis of tide gauge records, and more careful monitoring of reef flats is required to detect the expected recolonisation and spread of living corals. Where they develop and grow upwards to match sea level rise the outcome will be the formation of extensive and fragile 'reef gardens' rather than solid reef flats (Bird 1992).

Reef islands will be vulnerable to quite a small sea level rise, but if the next few decades see a strong revival of coral growth, this could afford them some protection. While a global sea level rise will have many impacts on the world's coastlines, coral reefs and reef islands are likely to be among the ecosystems that will be most strongly affected by world-wide submergence.

ARTIFICIAL AND DEVELOPED COASTS

Parts of the world's coastline have been modified in the course of development, and are now partly or wholly artificial. These include areas where ports have been constructed, together with breakwaters, canals, roads, railways and urban seafronts. Structures frequently encountered on the coast include sea walls and groynes, power stations, lighthouses, marinas, sewage treatment plants, outfall pipes and salt pans. In south-east Asia large areas of low-lying coast have been modified for the development of fish and prawn ponds, which are enclosed from the sea by low artificial banks, with or without a residual mangrove fringe.

There are many places around the world's coastline where developed property, including buildings, roads, farmed land and forestry areas, is threatened, or has been damaged or destroyed as a result of cliff recession or the erosion of beaches, deltas, salt marshes and mangrove swamps. Large sums of money have been spent by governments and local authorities on the building and enlarging of concrete sea walls and boulder ramparts where erosion threatened developed land in coastal urban areas, especially where beach resorts and tourist facilities have been constructed behind receding cliffs (Fig. 55) or on low-lying sandy coasts (Walker 1988). Artificial coasts have also been formed where land reclamation has required the building of an enclosing sea wall. Some reclaimed areas are polders, standing below high tide level behind a protective wall, as in The Netherlands, and maintained by drainage to low tide and by pumping off

Figure 55 Hotels built on coastal headlands near Pattaya, Thailand, are threatened by cliff erosion, which will worsen as sea level rises. A sea wall has been built to reduce basal erosion, but the cliff remains unstable

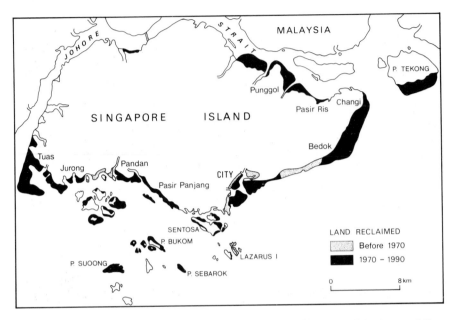

Figure 56   Land reclamation has been extensive around Singapore Island, especially since 1970, increasing the land area by about 10%. A sea level rise will necessitate the building up of these reclaimed areas, and it is likely that the threat of submergence will stimulate the reclamation of more areas from the sea. Based on information supplied by Dr Wong Poh Poh, University of Singapore

water; others have been built up by landfill behind a sea wall, as at Melaka in Malaysia, or simply by the dumping of sediment and rubble waste to advance the coastline, as in Singapore (Figs 56 and 57) (Wong Poh Poh 1985).

The prevalence of erosion on the world's beaches, mentioned previously, has prompted the building and elaboration of artificial structures designed to halt coastline recession. Solid sea walls and boulder ramparts usually cause wave reflection, which further depletes the beaches, and is particularly unwelcome at seaside resorts where the beaches were the original tourist attraction. An alternative method of shore protection is artificial beach nourishment, which has the advantage of providing or restoring a coastal recreational resource. Unfortunately it is expensive, and may be feasible only in a few intensively-developed urban resort areas (Schwartz and Bird 1990). As sea level rises it will be easier to nourish and maintain artificial beaches in coves and embayments than on long straight or gently-curving sandy beaches, which will require massive breakwaters to retain the sand within nourished compartments.

Where artificial structures have been built on the coast they have generally been designed in relation to present sea level. As sea level rises, existing sea walls and boulder ramparts will need to be raised and extended laterally, and

Figure 57   The southern coast of Singapore Island was advanced by extensive land reclamation in the 1970s. Segments of sea wall were built in the hope that the intervening beach sectors would become adjusted to prevailing wave patterns, and attain stability. However, erosion has continued, and the wall segments have been lengthened

new structures built to combat erosion and prevent the ingress of the sea on to developed agricultural, industrial or urbanised land (Stark 1988). This elaboration of sea walls has happened repeatedly along the subsiding coastline of The Netherlands over many centuries, and more than half the country is now polder land, below the level of high tides in the North Sea. Were it not for successive enlargement of protective sea walls, much of The Netherlands would have remained an estuarine wetland, and might by now have been designated a nature conservation area of global importance. Port structures will also have to be raised to maintain dock facilities for shipping.

A global sea level rise is likely to prompt demands for the enlargement and extension of existing sea walls as well as the building of many more, and the prospect is that long sectors of low-lying parts of the world's coastline will become as artificial as those of The Netherlands.

Chapter Four

Human responses to a rising sea level

INTRODUCTION

Chapter 3 indicated various scenarios for environmental changes likely to occur along the coast during the predicted global sea level rise. Human responses to such changes can be deduced theoretically, or from what has happened on subsiding coasts, where some kind of response has already been made to a rising sea level. There is much uncertainty, and the aim of this chapter is to provide a framework for considering the kinds of response that could occur, rather than to make firm predictions at this stage.

EVIDENCE FROM SUBSIDING COASTS

Sectors of the world's coastline known to have been subsiding in recent decades are shown in Fig. 2 (page 5). Within these, sea level rise has caused problems for local people, notably the flooding of low-lying areas and the losses of land through coastal erosion initiated or accelerated by submergence. The human response has varied with social, economic and political factors; it has been strong on urbanised and intensively developed coasts, and trivial where the coast is sparsely populated, or inhabited by primitive societies. This review will follow the numerical sequence shown in the caption to Fig. 2.

At Long Beach in southern California, oil extraction resulted in land subsidence of up to 9.5 m between 1926 and 1975 in the port and industrial zone west of the downtown area (Fig. 58). Buildings tilted, bridges were damaged, streets cracked, and underground pipes broke. The response included landfill operations, dock reconstruction, and the building and raising of dykes to confine marine inlets and the river mouth and so prevent flooding of low-lying areas (Fig. 59). After 1975 the subterranean oil reservoir was repressurised by

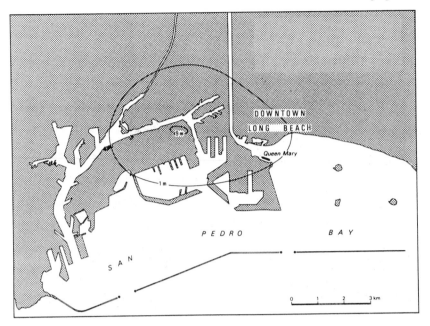

Figure 58   Land subsidence at Long Beach, California, as a result of oil extraction, attained more than 9 m between 1926 and 1975. It has been necessary to build riverside walls (Fig. 59) and raise the land area artificially to prevent marine submergence in the dock area to the west. Based on information provided by R. Stehle, City of Long Beach, and from Mayuga and Allen (1969)

pumping in salt water, and subsidence was brought under control. At Huntington, east of Long Beach, where land subsidence and nearshore deepening has occurred as a result of oil extraction, intensified wave attack resulted in the loss of the beach and an increased rate of recession of cliffs cut in soft sands and clays. The erosion has been halted by the emplacement of a large boulder rampart.

Sea level has been rising as a result of subsidence at the head of the Gulf of California, and has inundated areas of former saltflat and marshland. Only local efforts have been made to embank and protect settlement sites and low-lying farmland on the Colorado delta. The same is true of the swampy regions of the Parana delta at the head of the Rio de La Plata, north of Buenos Aires in Argentina, and on the mangrove-fringed islands and estuaries of the Amazon and Orinoco deltas.

On the Gulf and Atlantic coasts of the United States, land subsidence is thought to have worsened beach erosion and flooding from recurrent hurricane surges along many barrier island shores. Charleston (South Carolina) and Galveston (Texas) are among the coastal cities where large sea walls have been

Figure 59   Riverside walls keep the sea out of the subsided dock area west of Long
Beach, California. Note the active oil well to the right of the parked cars

built to protect urban areas from marine flooding. In Galveston flooding during
the 1900 hurricane led to the building of a sea wall in 1902–1905, and behind this
the land was artificially raised by forming dyked compartments, and pumping in
sand dredged from the sea floor. Between 1904 and 1910 over 2000 buildings
(including the opera house) were jacked up and underfilled, a procedure called
'raising the grade'. During the 1920s the nearby Gaillard Peninsula sank beneath
the waters of Galveston Bay as a result of oil extraction from the Goose
Creek oilfield. To the north, high sea walls have been built to prevent similar
marine flooding of the oil installations at Texas City, while at Brownwood,
near Houston, a housing estate built on subsiding land became submerged in
1975 (Davis, McCloy and Craig 1988). Subsidence due to underground sediment
compaction has lowered the land around Flushing Airport, New York, which is
now protected from sea flooding by dykes.

Many Gulf and Atlantic coast beaches have been maintained by artificial
nourishment. They include the beach in Harrison County, Mississippi, South
Seas Plantation on Captive Island, Florida, and Wrightsville Beach in North
Carolina. Behind the barrier islands, and on parts of the Mississippi delta,
embankments have been raised to keep tidal flooding out of farmland and
townships on low-lying ground behind salt marshes and reedswamp (called

roseau in Louisiana). There are, however, some areas bordering Chesapeake Bay where sea level rise has resulted in salt marsh plants invading farmland.

In southern and eastern England gradual subsidence has been accompanied by the erosion of cliffs, beaches and salt marshes. Sea defences built and extended over many centuries have endeavoured to halt land losses along cliffed coasts. Nevertheless, historical records and maps have been used to trace the disappearance of the port of Dunwich, and several other towns and villages along the east coast of England, where cliffs cut in soft glacial drift are exposed to the stormy North Sea, and have been receding by up to 200 m per century (Valentin 1971). Sea walls, groynes, boulder ramparts and other structures have been introduced, especially in urban areas, and extended stage by stage along the coastline. On the south coast of England, recession of chalk cliffs east of Brighton and sand and clay cliffs on the shores of Christchurch Bay has been halted by the building of undercliff promenades (Fig. 60). Beach depletion has become widespread, despite the building of numerous groynes intended to intercept and retain beach material. The increasing extent of artificial structures,

Figure 60   Cliff recession on the shores of Christchurch Bay, southern England, has been halted by the construction of a sea wall and esplanade, and the renourishment of the shingle beach. Cliff-crest recession continues, but the cliff is developing into a more gradually sloping bluff, which is expected to stabilise. As sea level rises it will be necessary to raise the level of the sea defences in order to prevent renewed erosion and further cliff retreat

especially along cliffed coasts of high scenic quality, as in Dorset, is causing concern, and there has been much talk of 'soft engineering', notably the use of artificial beaches as a means of halting cliff recession. In recent years beaches of shingle and sand have been emplaced at seaside resorts such as Bournemouth and Seaford.

Subsidence has resulted in increasingly high spring-tides and storm surges. Extensive marine flooding and erosion occurred during the North Sea storm surge in 1953, after which sea walls and estuary embankments were repaired and enlarged. The increasing risk of storm surge flooding in the London area led to the construction of the Thames Barrier at Woolwich (Fig. 61), designed to be closed when storm surges were moving up the Thames estuary, and so prevent inundation upriver (Gilbert and Horner 1984). Since it was completed in 1983 the Barrier has been used successfully to keep out surging tides, but these continue to be a problem downstream in the Thames estuary, where the tides may have been augmented by the various attempts to control sea flooding. Near Dartford a mobile gateway has been built to be lowered to keep Thames estuary storm surges out of the River Darent in north Kent (Fig. 62).

Figure 61   The Thames Barrier, completed in 1983, has rotating gates that can be raised to prevent storm surges raising the water level and flooding low-lying parts of London. Photograph reproduced by permission of the National Rivers Authority, Thames Region

Figure 62   On the Darent River, a tributary of the lower Thames, storm surge flooding
            can be excluded by lowering a gate to close the river channel

On the south coast of England, the Sussex village of Bosham, on the estuarine shores of Chichester Harbour, is flooded by high tides several times a year. Sea level rise, due mainly to land subsidence, now brings the highest tides into the village street (Fig. 63), and cottagers have had to raise the thresholds of gateways and doors to prevent the sea flooding into their homes (Fig. 64).

Along the southern shores of the Baltic Sea, cliffs cut in soft glacial drift have been receding rapidly, partly because land subsidence is causing a sea level rise. A few sectors have been armoured with sea walls and breakwaters, but on the Polish coast cliff recession has consumed substantial areas of farmland in recent centuries, and people have retreated inland as villages and farms have been destroyed by erosion.

The low-lying North Sea coast between Denmark and Belgium is fringed by the sandy Frisian Islands and the shallow Wadden Sea, where large areas of sand and mud are exposed at low tide. Medieval storm surges overwashed and destroyed extensive rural landscapes here, and the sea continues to rise as the land subsides. On the German island of Sylt, rapid recession of cliffs cut in soft glacial drift has prompted the construction of sea walls, and beaches have been renourished to protect the cliffs and maintain a recreational resource. Embankments (dykes) have been built and raised behind salt marshes to prevent high tide flooding of farmland and villages.

Figure 63 Bosham, on the south coast of England, is one of several seaside places subject to marine flooding during high tides and storm surges. The sea is shown invading the High Street. Marine flooding has become more frequent because of a gradual rise in sea level along the south coast of England, due partly to land subsidence

Sea walls have been built extensively in The Netherlands, where long-term subsidence has resulted in one-third of the country being below sea level (Hekstra 1986). Two thousand years ago the seaward half of The Netherlands consisted of dune-capped sandy barrier islands backed by tidal lagoons, salt marshes and freshwater swamps. Deposition of sediment, chiefly sand, silt and clay, was in progress, particularly in the south, around the mouths of the Rhine, the Scheldt and the Maas Rivers, where a delta was growing. In Roman times settlements were established on some of these barrier islands, notably at Domburg and Valkenburg, and Roman roads extended along the banks of rivers, notably the Eastern Scheldt, where a number of forts were built in the 1st century AD. Embanking of land, originally with dykes of earth and seaweed, began during the Roman era, but was on a very small scale until the medieval period, when successively more elaborate and extensive dykes, including timber pilings and earth-filled wooden walls, were constructed. Wooden structures suffered damage by epidemics of the pileworm *Toredo*, especially in the 16th and 17th centuries, and so stonework was increasingly used. In the past few decades traditional hand-lain stonework has been supplanted by large-scale mechanised construction of sea walls and sluices. Storm surge flooding from the North Sea

Figure 64   Cottagers in Bosham have adapted their doorways by raising the threshold
steps to keep out the recurring sea floods

in 1953 prompted the rebuilding of coastal dykes in a scheme of integrated sea
defences (De Ronde 1991).

The most widely publicised of these is the Dutch Delta Project (Fig. 65),
whereby the estuarine outlets from the Rhine, the Scheldt and the Maas have
been modified, and in some cases sealed, by massive new structures. This project
has reduced the length of the coastline of the southern Netherlands (the Dutch
'marine frontier') from 800 km to 80 km. It includes dams across the northern
outlets, and a storm surge barrier across the mouth of the Eastern Scheldt,
designed to close when sea level reaches above 3 m on the Amsterdam tide
gauge. The original scheme was to seal all of these openings, leaving only
the Western Scheldt estuary to the south, but Dutch environmentalists argued
successfully for the retention of the Eastern Scheldt as an estuarine system,
open to the ebb and flow of tides (albeit reduced by the sluices), because of
its ecological significance and importance for local fisheries. The Dutch Delta
Project was completed in 1986 at a cost of $US2000 million, and these structures
have provided an effective means of preventing sea flooding under existing sea
level conditions.

Other parts of The Netherlands coast have also been protected by large sea
walls. Westkapelle, at the western end of the island of Walcheren, once had a

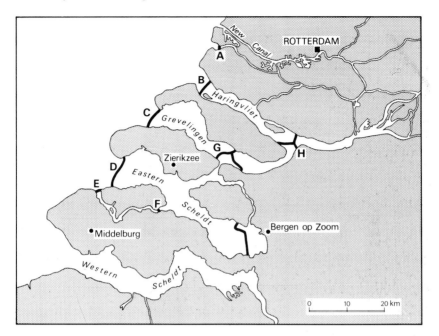

Figure 65   Following the disastrous sea floods that accompanied the 1953 North Sea storm surge, sea defences were repaired and raised on the coast of The Netherlands. In the Rhine delta area, the length of coastline to be defended was shortened by building a series of dams and sluices (A to H) to exclude storm surge flooding from river distributaries, and thus prevent any recurrence of marine flooding of Dutch lowlands. Based on information supplied by Delta Expo

dune fringe behind a sandy beach, but this was a point of divergence of longshore drifting, and as sand moved away both eastwards and southwards the beach was depleted and the dune fringe eroded. A sea wall originally built here in 1940 has been enlarged in several stages (Fig. 66). The present structure, completed in 1987, is 160 m wide and 10 m high (Fig. 67), with a seaward face consisting of a wide ramp of stonework, partly covered with tarmac, and accompanied by a series of groynes (Fig. 68). There is very little beach, even at low tide, and the coastal area looks as artificial as a motor racing track, in contrast with the sandy beaches and dunes that persist eastwards and southwards. Farther north, the low-lying coast at Den Helder is, like Westkapelle, protected by a high sea wall, and there are walls along much of the coast of the Wadden Sea, including parts of the inner shores of the islands of Texel, Terschelling and Ameland.

Between The Hague and Camperduin, the Dutch coastline consists of a beach backed by wide sand dunes. There has been erosion, especially during the past century, and the response in some sectors has been to armour the cliffed dunes with sea walls or boulders.

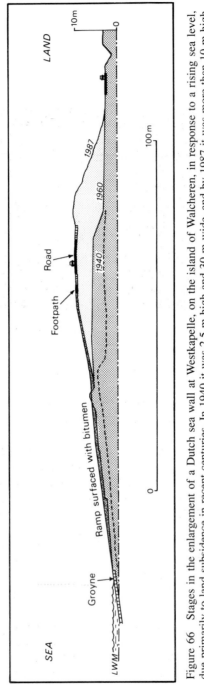

Figure 66 Stages in the enlargement of a Dutch sea wall at Westkapelle, on the island of Walcheren, in response to a rising sea level, due primarily to land subsidence in recent centuries. In 1940 it was 2.5 m high and 30 m wide, and by 1987 it was more than 10 m high and over 130 m wide, its seaward slope faced with stones and bitumen. LWT, low water mark. Based on information supplied by Delta Expo

Figure 67   The sea wall at Westkapelle after its enlargement in 1987

Similar responses have occurred on a smaller scale on subsiding parts of the west coast of France, notably around the Golfe du Morbihan and to the south of the Loire estuary (Miossec 1991), and near Lisbon in Portugal, where coastal towns and low-lying farmland and villages have been protected by embankments and shore defences.

On the Guadalquivir delta, and on several Mediterranean deltas, notably those of the Ebro, Rhône, Po, Danube and Nile, subsidence has led to submergence and erosion of low-lying land. Some areas have been embanked to protect farmland, villages and industrial zones, but considerable areas of marsh and alluvial land have been lost. On the Rhône delta, beach nourishment has been used to protect low-lying coast near Saintes-Maries-de-la-Mer.

Subsidence accelerated by extraction of underground water in north-eastern Italy has resulted in submergence and erosion of low-lying land. In response, sea defence works have been built along the Adriatic coast and embankments constructed to prevent marine flooding in Ravenna (Cencini et al. 1979). Beach erosion along the barriers seaward of the Venice Lagoon has been met by sea wall construction and the dumping of boulders, so that some beaches are now blocky heaps (Fig. 69) (Zunica 1990). Around the lagoon, embankments have been built to keep the sea out of farmland and to enclose fishponds. The worsening problem of sea flooding ('aqua alta') in the city of Venice is to be countered

Figure 68   The seaward slope of the sea wall at Westkapelle forms a stable, but artificial, coastline. Wooden posts form rough groynes, intended to trap sand and form a beach, but there is very little beach material here

by the insertion of barrages to close the three marine entrances to the Venice Lagoon during storm surges, much in the way that the Thames Barrier keeps marine flooding out of London (see page 107).

Marshlands have been submerged on the eastern shores of the Sea of Azov and along the seaward fringe of Poti Swamp on the Georgian Black Sea coast, where artificial beaches have been used to halt the erosion of salt marshes. In Africa, erosion of subsiding parts of the Niger and Zambezi deltas has met with only minor efforts to preserve or reclaim submerging land areas. The Tigris–Euphrates delta has embanked areas, notably on the island of Bubiyan, but there has been considerable damage during recent wars. In the Rann of Kutch low-lying alluvial land has been submerged by the rising sea, and there has been recurrent storm surge flooding in south-eastern India and on the subsiding Ganges–Brahmaputra delta. Sea flooding on the coast of Bangladesh has prompted extensive building of earth embankments to protect densely-populated ricefield and fishpond areas, but many of these have been breached and overwashed in recent storm surges.

In south-east Asia land has been lost as a result of sea level rise on submerging parts of the Irrawaddy, Mekong and Red River deltas, but in many places embankments have been built to keep the sea out of ricefields and villages.

Figure 69   Sea level rise resulting largely from land subsidence in the Venice region has led to beach erosion on the enclosing barrier islands. Here at Litorale di Pellestrina the lost beach has been replaced by a sea wall and boulder rampart

Subsidence in the Bangkok region as a result of groundwater extraction has made river flooding more extensive, but has affected only a short sector of coastline (Fig. 9, page 20), where embankments have been built behind the mangrove fringe to keep the sea out of ricefields and fishponds (Somboon 1990). In Japan, Niigata and Maizuru are coastal towns much affected by subsidence. At Niigata (Fig. 70) a large sea wall has been built, and the former beach in front of it is covered with concrete tetrapods (Fig. 71). The disposal of floodwater from rain and rivers has become a problem here, because much of the urban area is now below sea level and gravitational drainage is no longer possible. Pumping has been introduced to remove surplus water (Fig. 72). Similar responses have occurred in the Taipeh region of northern Taiwan, where subsidence of up to 2 m between 1955 and 1977, induced by groundwater extraction, led to problems of draining floodwaters from the Tamsui River (Chin Kuo 1986).

Submergence and erosion on the shores of the Straits of Melaka have resulted in losses of mangroves and some reclaimed land in eastern Sumatra and West Johore. In places banks have been built to keep out the advancing sea (Economic Planning Unit (Malaysia) 1985). On the deltaic coast of northern Java erosion and submergence have destroyed brackish-water fishponds (Fig. 73) and salt penetration has damaged the ricefields that lie behind them (Fig. 74). People who

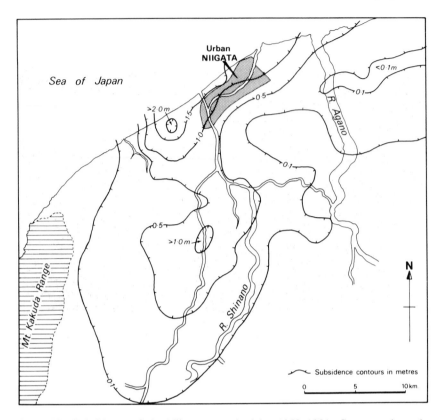

Figure 70   Subsidence of the Niigata coastal plain, 1959–1981. Contours show the pattern of land subsidence, due largely to groundwater extraction, recorded in the region of Niigata, on the north coast of Honshu, Japan, between 1959 and 1981. Based on information provided by K. Koike, Komazawa University, Tokyo

lived in the coastal fringe and operated these have moved away. On the shores of the Sepik delta, in New Guinea, repeated overwash by storm surges has driven a sandy barrier landward (Fig. 75), and the response of native villagers has been to move back and rebuild their settlements after each storm surge (Bird 1986). Rapid erosion of the old delta of the Hwang Ho, in northern China, abandoned when the river was diverted northward in a flood in 1855, has been halted by extensive sea walls. *Spartina* grass has been introduced to mudflat areas to diminish wave attack on these structures and help to conserve the coastline (Chen Jiyu, Lu Cangzi and Yu Zhiyang 1985).

   In the Port Adelaide region, where subsidence has led to the landward retreat of mangrove and salt marsh vegetation (Burton 1982), the response has been to dump earth and rubble on low-lying Garden Island as a prelude to industrial and

Figure 71   Land subsidence at Niigata, Japan, indicated in Fig. 70, resulted in the disappearance of the beach and the subsequent armouring of the coastline with large concrete tetrapods and breakwaters

marina development, and the building of a Multi-function Polis. This could be seen as a form of counter-attack. Corner Inlet, in Victoria, is another subsiding embayment fringed by mangroves, salt marshes and some beaches, backed by a rural landscape. Embankments have been built to enclose farmland, but the area is sparsely populated and there has been little response to erosion as sea level rises, except for the building of a stone wall on a small sector near Foster after earlier brushwood fencing had failed to halt erosion.

The sea level rise of about 1.5 m in the Caspian since 1977 has forced the abandonment of extensive land areas that had emerged during the preceding regression between 1930 and 1977, when about 30 000 km² of former sea floor were exposed. About half of this was developed for agriculture, with 200 new villages, and 1000 km of roadways. Some urban settlements expanded seaward, and the rising sea has flooded into these, notably in the Azerbaijan city of Astara. In all, more than a million people have been affected by the sea level rise, and the losses since 1977 are about 20 million roubles (at 1991 values). In the Russian sector, there are plans to build a coastal highway protected by a sea wall.

Figure 72   Pumping stations are necessary to keep low-lying parts of urban Niigata (Fig. 70) free from marine flooding and to remove floodwaters from the Shinano River

This brief review of human responses to a rising sea level on sectors of subsiding coast provides a background for considering how people will react to a global sea level rise.

RESPONSES TO THE PREDICTED SEA LEVEL RISE

During the past few years there have been numerous attempts to predict human responses to the changes that are expected as sea level rises, notably in countries that will be strongly affected, such as the Netherlands, Egypt, Bangladesh, and oceanic island states (e.g. De Ronde 1991, Wind 1987). The Maldives, for example, are a nation in the Indian Ocean consisting of a group of atolls with more than 1000 scattered low islands, 200 of which are inhabited by a total population of about 180 000 people. They have been the subject of numerous reports (e.g. Edwards 1989, Hulsbergen and Schröder 1989, Pernetta and Sestini 1989, Woodroffe 1989). There has also been discussion of the responses to sea level rise in subsiding areas such as Venice and Bangkok.

Chapter 3 showed the difficulty of predicting the extent to which coastal environments will change as a result of a sea level rise of about 1 m. While

Figure 73   Destruction of brackish-water fishponds on the western shores of the Cimanuk delta, northern Java, as a result of a sea level rise due largely to subsidence of a deltaic area no longer maintained by fluvial sedimentation

it is relatively easy to map submergence to a specified contour, there are large margins of error in predicting the extent of accompanying erosion and accretion, and the gains and losses of land area that will result. Some attempts have nevertheless been made to predict the extent of coastline retreat, with maps that show the areas likely to be submerged and consumed by erosion as sea level rises. In Britain Boorman, Goss-Custard and McGrorty (1989) produced a map showing the land areas that would be submerged by a sea level rise (Fig. 76). A similar map of the area that would be submerged by a 1-m sea level rise in the Bangkok region, at the head of the Gulf of Thailand (Fig. 77), was presented by Bird (1989c).

Responses to the predicted sea level rise in Port Phillip Bay, Australia, were discussed by Bird (1988c). Estimates of the extent of submergence as a result of a 1-m sea level rise were made for each of the various kinds of coast that border this bay: cliffs, bluffs and rocky shores with or without beaches; sandy shores backed by beach ridges; salt marshes and mangroves; alluvial plains and sectors fringed by sea walls or boulder ramparts. On the beach-fringed sectors the rule-of-thumb proposed by Bruun (1962) was used as a first approximation, i.e. that a 1-m sea level rise would lead to coastline recession by up to 100 m, and

Figure 74   Damage to ricefields by salt penetration behind the fishpond zone on the western Cimanuk delta, northern Java, as a consequence of sea level rise and coastal recession (cf. Fig. 35, page 68)

(in the absence of alternative models) similar retreat was envisaged on shores where the rising sea will encounter soft sediments, including marshland and alluvial plains (Fig. 78). The extent to which these changes will be offset by sediment accretion during submergence is likely to be very limited, because the few inflowing rivers are carrying only meagre loads, mostly silt and clay rather than sand, and the nearshore shallows that could become sources of potential shoreward sediment drifting are not extensive.

Prediction of responses to sea level rise on a regional basis around the Australian coastline were explored with the aid of the computer-based Australian Resources Information System by Cocks, Gilmour and Wood (1988). In Papua New Guinea, Sullivan and Hughes (1989) prepared maps to show the expected extent of severe inundation, defined as coastline retreat of more than 500 m or losses of more than 1 km^2 of land, and moderate inundation, where the losses would be less than this. Gornitz and Kanciruk (1989) used a Geographical Information System to screen out high risk areas in terms of sea level rise around the United States. Such areas had characteristics such as low-lying, erodible substrates, active subsidence, shorelines already in retreat, and exposure to high wave and tide energy. Chesapeake Bay, the North Carolina Outer Banks,

Figure 75   On the north coast of the Sepik delta, New Guinea, villages built on a low sandy barrier are destroyed by recurrent storm surges, which drive the barrier landward. As the coastline is set back, the villagers retreat and rebuild their homes

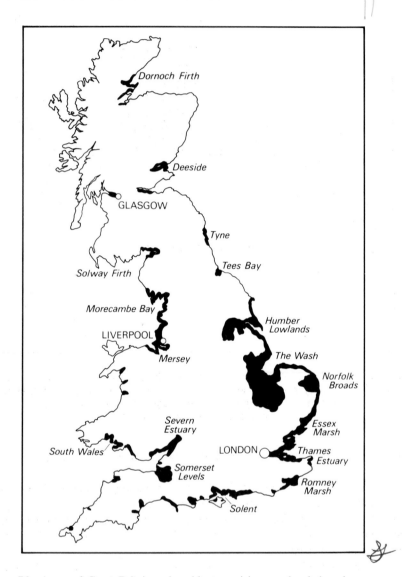

Figure 76   Areas of Great Britain vulnerable to a rising sea level, based on a map published by Boorman, Goss-Custard and McGrorty (1989). A sea level rise of several metres would be necessary to inundate these areas

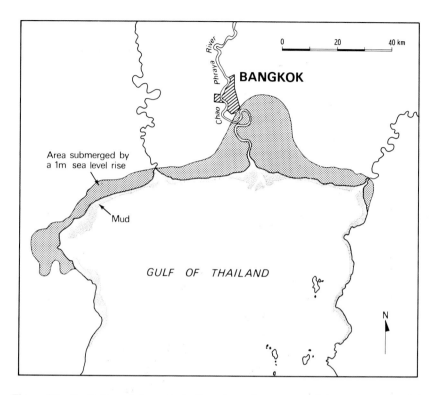

Figure 77   Prediction of the extent of marine submergence of the coast north of the Gulf of Thailand by a 1-m sea level rise, if no walls are built to prevent it. Note the landward protrusion east of Bangkok where submergence will extend into an area of land subsidence (see Fig. 9, page 20)

southern Florida, and the central Gulf coast were identified as major high risk areas.

Predictions such as these can be refined as modelling is improved, and tested by monitoring coastal changes in the early stages of a sea level rise (e.g. to 10 cm). This will provide a basis for subsequent extrapolation, and the shaping of strategies for coastal management and development around the world's coastline through the 21st century.

Coastal submergence by up to 1 m over the next 100–150 years is difficult to envisage, because it is considerably greater than that recorded on the world's gradually subsiding coasts during the past century. Goudie (1986) quoted reports of submergence of parts of the Black Sea coast at rates of up to 52.5 cm per century, and up to 30 cm per century from parts of Indonesia. The land is subsiding in The Netherlands at up to 20 cm per century, and on the south and

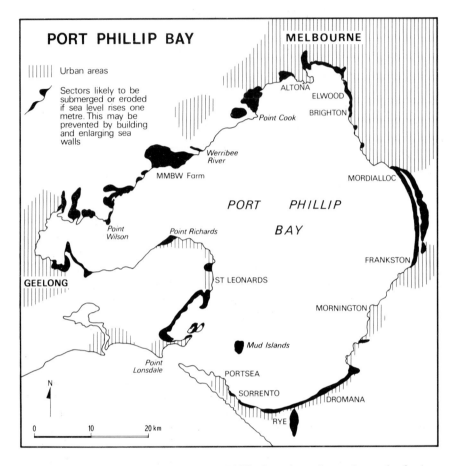

Figure 78   Changes expected around Port Phillip Bay, Australia, as the result of a 1-m
sea level rise

east coasts of England at up to 10 cm per century. Subsidence in the Venice
region has resulted in a sea level rise of 30 cm during the past century.

Sudden submergence of more than 1 m has occurred on some coasts as
a result of earthquakes. These events have been localised, usually with rapid
ensuing erosion and landslides on steep coasts. The effects of the 1964 Alaskan
earthquake in the Anchorage area are well documented (Shepard and Wanless
1971). In some places buildings and other structures have been damaged and
people have abandoned land areas that became permanently flooded as a result
of earthquake subsidence.

HUMAN RESPONSE SCENARIOS

The ways in which people have responded to changes on subsiding coasts have varied with social, political and economic factors. A global sea level rise is likely to produce three kinds of response:

(1) Evacuation, abandoning the land as it is submerged and eroded, and where possible adapting to the effects of the rising sea
(2) Attempts to maintain the existing coastline and coastal margin by engineering works
(3) Counter-attack, notably by building walls to enclose and reclaim intertidal and nearshore areas

Societies that have the organisation, technology, and resources to counter the effects of a sea level rise by means of anti-erosion structures and land reclamation will do so, following the example of The Netherlands; others will in general have to evacuate submerging areas and adapt their activities and economies to changing coastal environments. Possible responses will be considered in terms of the three scenarios, the discussion following the sequence of environmental features used in Chapter 3.

Scenario 1: Evacuate and adapt

Losses of coastal land have occurred as a result of cliff recession, beach erosion, and the retreat of sectors of shoreline on deltas and coastal plains. Apart from urban and industrial areas, and ports and seaside resorts, where sea walls and other structures have been built to halt coastline recession, the human response has been to abandon the threatened structures (Figs 79 and 80) and retreat from the disappearing land. Even where the coastal fringe has been developed, abandonment may be the most economic strategy if the cost of protecting the existing coastline exceeds the value of the structures that are threatened by erosion and submergence. Few buildings, roads and bridges have been designed to last more than 50 years without substantial renovation, and the cost of maintaining them has to be considered in relation to the expense of maintaining the terrain they stand upon.

Erosion and submergence of beaches as sea level rises will be a major problem in areas that have been developed as seaside resorts. Where the beaches fringe bluffs and cliffs they may disappear altogether, unless artificially replaced; where they border beach-ridge plains or are backed by coastal dunes, beaches will continue to exist as the coastline retreats through the sandy terrain, providing that solid structures such as sea walls and promenades have not been built to prevent this. Where seaside resorts have developed the retreat option will

Figure 79   Cliff recession has forced the abandonment of this house on the coast of
Christchurch Bay, southern England

mean moving or abandoning structures that have been built for recreational
use and tourist accommodation. Shore features such as marinas, boat ramps,
beach huts and boatsheds can be rebuilt at higher levels as the sea rises. On
the coastal fringe mobile features, such as caravan grounds and camping sites,
can be moved back landward, but where more solid structures such as hotels,
shops and other buildings have been emplaced there will be arguments against
abandoning them. Where the coastal development consists only of small and
relatively inexpensive buildings, they can more easily be abandoned. At Dutton
Way, in Victoria, Australia, for example, local authorities decided to abandon
a seaside bungalow development because of the high cost of protecting it from
coastline erosion, and there are many coastal shack towns that would not be
maintained in the face of a rising sea level. Some coastal management agencies
have already adopted set-back lines, seaward of which development must take
account of the possibility of land being lost by erosion. The implication is that
these lines should now be drawn further landward to allow for the effects of a
rising sea level as well as erosion.

On low-lying coasts, including deltas, alluvial and beach-ridge plains, barriers
and barrier islands, submergence by the rising sea will be accompanied by
erosion, and the coastline may recede by hundreds of metres to several

Figure 80   Beach erosion on Rhode Island, United States, has damaged and destroyed holiday shacks built on a barrier island

kilometres. On parts of deltas already showing erosion because of the reduction of sediment yield by dams, as on the Nile delta, local inhabitants have already abandoned farms and villages, and structures such as fishponds and salt pans. It has been estimated that up to 20% of the Nile delta would be lost and evacuated as the result of a 1-m sea level rise, the coastline retreating by up to 30 km; between 8 and 10 million people would be displaced, and up to 15% of Egypt's arable farmland destroyed (Delft Hydraulics Institute 1991, Milliman, Broadus and Gable 1989, Sestini 1991).

Similar changes have occurred where erosion and inundation have resulted from land subsidence, as on the Mississippi delta. Although long-term subsidence has been gradual, the effects are felt when flooding and erosion occur during major storms, when strong winds raise sea level and drive large waves on to the fragile coast. Destruction is then widespread, and damaged areas may have to be abandoned as a result.

An early example of this was the North Sea storm surge of 1362, which is said to have forced the abandonment of a large area of low-lying and subsiding agricultural land that lay to the west of Schleswig-Holstein in Germany (Davies 1972). This is now a broad intertidal and nearshore sea area backing the Heligoland Bight, with areas of slightly higher ground persisting as islands such as Sylt, Amrum, Pellworm and Buschsand. Shoals of sand and mud move to and fro on the intervening shallows, at times uncovering relics of field patterns and settlement sites.

More generally, the ancient Greek legend of Atlantis (a lost island west of the Strait of Gibraltar: there are equivalents in other historical cultures) may recall, through folk memory, the effects of the Late Quaternary marine transgression. As has been noted, this global sea level rise took place between 18 000 and 6000 years ago, at an average rate of just over 1 m per century, submerging land fringes as continental shelves (Fairbridge 1961). There must have been a long phase during which prehistoric people were obliged to retreat from the submerging landscapes. The consequences may be imagined in such areas as the Isles of Scilly, off south-west England, which are a partially submerged (and still subsiding) granitic landscape forming an archipelago with a number of sea floor features that have been interpreted as drowned buildings and field boundaries, perhaps similar to the Stone Age hut circles and associated structures still to be seen on the uplands of Dartmoor and Bodmin Moor.

Prehistoric communities that lived in coastal lowlands were forced to retreat as the Late Quaternary marine transgression took place. Among these were the Australian aborigines who, 18 000 years ago, when sea level was at least 140 m below its present level, were able to walk to and fro across coastal lowlands between Papua and Australia, and between mainland Australia and Tasmania (Fig. 81). As the sea rose it submerged these 'land bridges' to form Torres Strait (Walker 1972) and Bass Strait (Jennings 1971). The aborigines of Tasmania became separated from those of the mainland about 12 000 years ago, when the rising sea flowed across the land isthmus south of Wilsons Promontory to revive Bass Strait, which then became wider and deeper until it attained its present configuration about 6000 years ago (Bird 1987).

In recent decades, people have been retreating from low-lying coastal areas where erosion has been rapid. On the north coast of Java, for example, sectors of deltaic coastline have retreated by as much as 500 m in the past three decades. The land lost was partly mangrove swamp, but also included substantial areas in the upper intertidal zone where local people had converted mangrove areas into brackish-water fish and prawn ponds. The simplest of these are the traditional *tambak*, embanked areas with sluices to permit the gravitational inflow of sea water at high tide, and draining for harvesting at low tide, which have existed for many centuries on the north coast of Java. In recent years the demand for new fishponds has been so strong that they have been constructed on mud deposited at river mouths by floodwaters as soon as the floods disperse, and before mangroves are able to colonise (Bird and Ongkosongo 1980). Where marine erosion has breached their enclosing banks these fish and prawn ponds now lie derelict (Fig. 73). Landward of the tambak are ricefields, and as erosion proceeded these have been damaged by saltwater intrusion. People who occupied and used this

Figure 81   The outline of Australia and the islands to the north about 18 000 years ago, prior to the Late Quaternary marine transgression. Sea level then rose to submerge the bordering continental shelf, and separate Australia from New Guinea by the formation of Torres Strait (about 12 000 years ago), and from Tasmania by the formation of Bass Strait (about 9000 years ago)

coastal terrain have either retreated inland to higher ground or moved to other coastal areas as it became damaged and eroded. In some places the landward spread of salt water has already prompted adaptation, in the form of conversion of ricefields to brackish-water fishponds. Such changes have raised problems for rice farmers, who are losing land and resources as areas are converted to brackish-water fishponds.

A similar situation occurs extensively along other parts of the coast of south-east Asia. Coastal lowlands have been developed into a typical zonation of nearshore shallow sea, intertidal mudflats, residual mangrove fringe, embankment delimiting high tides, then brackish-water fish or prawn ponds in the zone permitting gravity drainage and irrigation by sea water, ricefields irrigated by fresh water at slightly higher levels, and a rising hinterland with forest, plantations or dryland farming (Fig. 82a). Such a zonation recurs, with minor variations, along many sectors of the Thai, Indonesian and Philippines coast, and to a lesser extent in Malaysia, where the fish and prawn ponds are still relatively localised. Maintenance of such systems as sea level rises will require expensive engineering works, and in some areas it may be more practical and economic to retreat and adapt. If the seaward fringe of brackish-water fishponds is abandoned it will be replaced by landward-migrating intertidal mangrove and mudflat areas, which have value as nursery and feeding grounds for fish and shellfish. The ricefields immediately to landward can be converted into brackish-water fishponds, and new irrigated areas can be established for rice production at higher levels in the hinterland (Fig. 82b). It would not be difficult to convert rice fields into fish or prawn ponds as sea level rises, but there are questions of land tenure and social equity. In practice, it could be the rice farmers in the immediate hinterland who are most disadvantaged, if pond operators move back to take over the ricefields for aquaculture. The area farmed for rice could be reduced, and rice production from the coastal regions could fall, in order that fish and prawn production may be maintained (Bird and Missen 1990).

More elaborate ponds, developed mainly for tiger prawn production in recent decades, have displaced large areas of mangroves in south-east Asia. Tiger prawn farming was initiated in Taiwan, in areas enclosed by sea walls, with pumping systems to maintain the sea water supply and the use of aerators, breeding techniques and fertilisers to generate high productivity from intensive aquaculture. As sea level rises, these are less likely to be abandoned than the primitive intertidal *tambak*, but their seaward walls will have to be fortified against increasingly high tides and strengthening wave attack. There may be other difficulties, for the disease risk is high in these intensive tropical aquaculture systems, and they may be vulnerable to the increasing ultraviolet radiation that results from depletion of stratospheric ozone. South-west of Bangkok brackish-water prawn ponds, abandoned because of recurrent disease problems, are being recolonised by mangroves.

In south-east Asia extensive areas of low-lying coastal plain are protected by banks of earth and stone, built along the landward boundary of a mangrove fringe to protect land developed for plantation agriculture, notably coconut and oil palm. In many places on the Malaysian shores of the Strait of Malacca the mangrove fringe has subsequently been cut back, and where the mangroves have been removed the embankments are exposed to wave attack. Some have been breached, and the sea has invaded plantations that lay behind them. Where embankments have been submerged and destroyed by erosion and the plantations abandoned, the relics of coconut trees may be seen on the muddy foreshore at low tide.

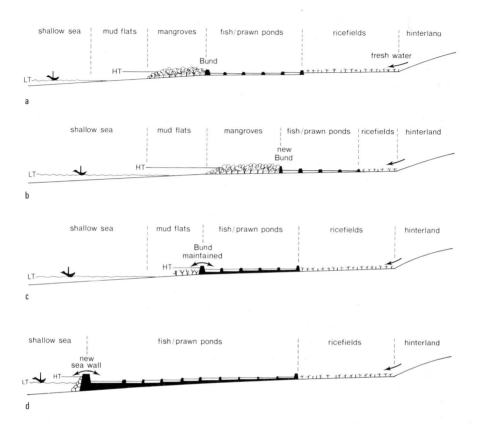

Figure 82   Scenarios for responses to sea level rise on the coast of south-east Asia. (a) Typical land-use zonation; (b) changes if the seaward margin is abandoned as sea level rises; (c) the situation if a sea wall is built to maintain the present coastline, the mangrove fringe being removed by erosion; (d) the situation if a sea wall is built offshore (the 'counter-attack' option) to prevent submergence by the rising sea, and the former mangrove and mudflat zones are reclaimed for productive use

As submergence proceeds there is a strong possibility, already in evidence in parts of the Philippines, that increasing use will be made of nearshore waters for fish and prawn farming in floating enclosures. Shallow sea areas may be occupied by people who live on boats, and who may eventually build floating villages and towns (Juliano, Anderson and Librero 1982). Problems of land tenure and social equity in coastal regions may then be overshadowed by problems of sea tenure over the submerging lands (Ruddle and Johannes 1985).

In the coastal lowlands of Bangladesh there is already a problem with the recurrence of disastrous marine flooding during storm surges in the Bay of

Bengal, a situation that will worsen as sea level rises. The densely populated lowlands consist of the deltaic plains of the Brahmaputra, Ganges and Meghna Rivers, and a sea level rise here would permanently inundate about 17% of the country (Fig. 83). If this area has to be abandoned, Bangladesh would lose about 8% of its gross domestic product (1984–1985), at a cost of about $US1 billion (Broadus et al. 1986). On the other hand, the maintenance and protection of a long, low, vulnerable coastline would require expenditure on a scale that is well beyond the means of Bangladesh.

As sea level rises, salt water and tidal movements will penetrate farther upstream in estuaries. In China, for example, salt water now runs 130 km up the Yangtze River, and a sea level rise will extend this, causing extensive salinisation as water spreads through interconnecting canals, and making ricefield irrigation impossible. It has already been necessary to relocate some freshwater wells. One

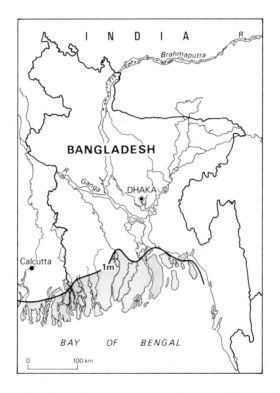

Figure 83   Extent of marine submergence of low-lying areas (shaded) in Bangladesh if sea level rises 1 m. Based on a map presented by Milliman et al. (1989)

option will be to abandon the salt-affected areas; another to convert them into forms of agriculture or aquaculture that can continue in brackish environments.

The effects of a sea level rise on marine fisheries will be minor off steep coasts where there is deep water inshore, but on lowland coasts with wide intertidal and nearshore zones there will be modifications as the various ecological zones migrate landward. Estuaries and coastal lagoons are important as breeding and feeding areas for fish and shellfish, and will remain productive where the sustaining ecosystems shift landward as sea level rises, but if these ecosystems are reduced by such changes management may be required to maintain their productivity. Some species of fish and shellfish may be favoured, others diminished, but in many places a landward transgression, submerging nutrient-rich intertidal and coastal soils, may lead to an increase in productivity of the migrating ecosystems and their accompanying fish and shellfish resources. People concerned with fishing and shellfish harvesting may have to change the location of their homes, villages and harbours as sea level rises to submerge sites they presently occupy, and adapt to the changing distribution of fishing and shellfishing areas. In so doing they may move back into areas that had previously sustained agriculture or forestry, and find they are competing with or displacing people whose livelihood had depended on the utilisation of land now submerged, salinised, or eroded; and these people, in turn, may have to retreat and adapt to the changing environment.

In tropical regions, where coastal lowlands are often densely populated, a sea level rise will displace large numbers of people. In south-east Asia it has been estimated that a 1-m sea level rise will submerge and erode some 5000 km^2 of land in Thailand, a similar area in Malaysia, and at least twice this area in Indonesia. Over the three countries several million people will be affected, posing social, economic and political problems of large-scale resettlement comparable with those that have arisen in the course of the Indonesian transmigration programmes, in which people from overcrowded Java have been resettled in Sumatra, Irian Jaya, and elsewhere.

It seems probable that mangrove areas, already much reduced by various human activities and impacts, will decline further as sea level rises, especially where embankments built along their landward margins prevent them spreading back into the hinterland. In south-east Asia a large number of people utilise mangrove ecosystems. In Thailand, for example, about 98 000 people live in or near the mangroves, mostly at the landward edge. There are many fishing villages within the mangrove area and on outlying mangrove islands (Kunstadter, Sabasri and Bird 1986). There are also mangrove villages in western Malaysia. Around the Segara Anakan, an estuarine lagoon in southern Java, about 8000 people live in the mangrove area, moving their villages seaward as the mangroves spread on to accreting tidal mudflats. Mangroves have sometimes been regarded merely as waste areas awaiting reclamation for productive use, but they are ecologically very rich. They are important as a breeding area for fish and prawns; they sustain important fisheries; they protect the shore from erosion and promote accretion by trapping drifting sediment; and they are used for the production of charcoal, firewood, poles, timber and fishing gear, as well as incidental food (crabs and edible fruit) and some medicinal aids. Even where people do not actually live in mangrove areas, neighbouring communities make use of them by frequently visiting them in search of these various products.

Already disadvantaged by the reduction of the mangrove area in recent decades, these people will have to modify their activities and move their settlement sites as sea level rises. Where the mangroves are able to migrate landward, their occupants can follow; people who were living on bordering land can relocate to higher levels, and existing uses of the mangrove area can persist as it shifts. Villages actually in the mangroves can be moved back in response to sea level rise as easily as they moved forward in the Segara Anakan in response to changes accompanying accretion, at least until the mangrove fringe is extinguished. If the mangroves diminish, and eventually disappear, the resources they now provide will also be depleted, and the people who used them will have to seek alternatives.

The persistence of coastal wetlands as nature conservation areas requires that they can migrate landwards as sea level rises. In Britain, for example, the prospect is that salt marshes and tidal mudflats will be much reduced by a rising sea (Boorman, Goss-Custard and McGrorty 1989). Many former wetland areas have been embanked and reclaimed for agricultural use, but the farmland is now often marginal economically. Sea walls built a century or more ago now require repair and reconstruction, and in some places it may be deemed preferable to abandon them, allowing the sea to invade and re-create salt marshes and tidal mudflats as sea level rises. There are already wetland nature reserves at Walberswick in Suffolk and Hamford in Essex where farmland previously embanked has been reopened to tidal flooding, and there are many other areas where the 'retreat option' may be taken as sea level rises (Brooke 1991, Doody 1991). Some fourteen salt marsh areas have been re-created in this way in a Tidelands Restoration Project on the shores of San Francisco Bay. Titus (1991) has explored the social, legal, economic and planning implications of allowing coastal wetlands to transgress landwards in the United States: their persistence in this manner would require abandonment of existing land uses in an area the size of the state of Massachusetts.

Evacuation and abandonment as sea level rises seems in prospect for many small, low oceanic reef islands, such as the Maldives in the Indian Ocean and Kiribati in the mid-Pacific. Unfortunately, there were many new tourist developments in the Maldives in the 1980s, and their maintenance poses an increasing problem. As has been noted, a sea level rise is likely to submerge and overwash low sandy cays on reefs and atolls, and cause upward movement, with depletion and salinisation of their groundwater resources. These are lenticular freshwater bodies within the low island sediments, upon which natural vegetation and crops depend, and they will shrink as they are forced upwards by the rising sea level; at the same time intruding sea water will make them more saline. These changes will modify the vegetation cover, and the ensuing ecological changes will be adverse for human occupance and agriculture. These are matters of great concern in the Maldives (Pernetta and Sestini 1989) and on other oceanic islands, especially in the mid-Pacific (Nunn 1990).

The migration of people from eroding and submerging coastal land areas inland, or to other coastal sites, will raise many social, economic and political problems. Considerable changes are likely to occur in land use and settlement patterns in the immediate hinterland, both because of rising water tables and the risk of increasing salinisation as sea level rises, and because of competition from people retreating from the coast. On deltas and coastal plains there may be changes in land use, for example a conversion from agriculture to aquaculture, or intensified crop production, or increased urbanisation and industrial development. Areas that now produce and export food may need to start importing it. Even with an evacuate and adapt scenario, the changes that actually occur in population, land use and economic development in most coastal regions will be as strongly influenced by social, economic and political considerations on regional, national and international scales as by the environmental changes accompanying a sea level rise.

Scenario 2: Hold the coastline

The most likely human response to a sea level rise, especially in long-settled urbanised areas, will be to try to maintain the present coastline by constructing sea walls, and to introduce drainage systems to prevent submergence of low-lying coastal land, and to dispose of rain and river floodwaters. This could be called the 'Dutch solution'. In The Netherlands, where coastal subsidence has prompted the building and enlargement of sea defences over many centuries, about half the coastline is now artificial, consisting of sea walls. The aim now is to hold the 1990 coastline by raising sea walls and renourishing sandy beaches in front of eroding dunes (De Ronde 1991). It has been estimated that the cost of maintaining Dutch coastal sea defences as sea level rises 1 m will be $US12 billion (Peerbolte, De Ronde and Baarse 1991). The cost of maintaining the present coastline of the United States in this way would be at least $US500 billion. In general, it is likely that only the most heavily urbanised and densely populated coastal lowlands will be protected in this way, perhaps with some areas of highly productive agriculture and aquaculture, and possibly sites of military significance.

Various attempts have been made to halt cliff recession, and so prevent further losses of cliff-top land, especially in areas where urban or industrial development has taken place. The usual procedure is to build a cliff-base wall or boulder rampart, sometimes providing an undercliff walkway or promenade, as in many seaside resorts. The sea wall is intended to prevent waves attacking the cliff base, and when the rock outcrops are coherent and relatively resistant, as in the English chalk cliffs, such basal protection is sufficient to stabilise the cliff, apart from recurrent rock falls from the cliff face and slight recession of the cliff crest. On softer formations, slumping and subaerial weathering and erosion of the cliff continue, degrading the profile to a gentler, more stable slope, but stability is attained only after considerable cliff-crest recession and loss of land. As sea level rises, overtopping of existing sea walls and boulder ramparts by storm waves will revive cliff-base erosion, undercutting the cliff face and causing renewed recession. If the coastline is to be maintained, cliff-base walls will have to be enlarged to higher levels to prevent this happening.

Beach erosion can be prevented, or at least slowed down, by building nearshore or submerged breakwaters to intercept wave action before it reaches the shore. Such structures are more effective where the tide range is small, and many sectors of Mediterranean beach-fringed coast have been stabilised in this way. Stability can be maintained as sea level rises by building these structures upward. When beaches are being cut back into sandy or gravelly terrain, such as beach-ridge plains, barriers and barrier islands, or coasts backed by dune formations, attempts to stabilise the coastline by sea wall construction can result in scour that deepens nearshore water and removes the beach from their seaward side.

Beach resorts and tourist facilities have been developed extensively on low-lying sandy coasts. In many places, hotels have been built on beach-ridge plains, and these will be threatened as the coastline is cut back by a rising sea. Structural works such as concrete sea walls and boulder ramparts are likely to be built and enlarged to protect developed seaside land from submergence and erosion.

The spread and acceleration of beach erosion as sea level rises are likely to be met by the proliferation and elaboration of artificial structures such as the heaps of tetrapods that now disfigure long sectors of the coastline of Japan (Walker 1988), even though it is now clear that such structures promote wave reflection scour, which depletes eroding beaches still further. In the United States, it has been estimated that the building and maintenance of such prevention works now cost at least $US300 000 per kilometre. Here, and in other western countries, the problem will be seen mainly in terms of insurance costs (e.g. through the American National Flood Insurance Program), legislation, compensation, and planning controversies.

Beaches that are valued for recreation or tourism, as in seaside resorts, may be maintained by beach renourishment programmes of the kind already used in the United States and Australia. Between 1962 and 1985 the United States Army Corps of Engineers renourished over 700 km of beaches, using 1.3 billion m^3 of sand, at a total cost of $US8 billion (Schwartz and Bird 1990). Such operations may be economic and feasible only as sea level rises in a few intensively-developed urban resort areas, such as the developed barrier islands on the Gulf and Atlantic coasts of the United States (Platt, Warrick and Wigley 1983), but in Britain there are plans to protect a 14-km sector of The Wash coastline between Heacham and Snettisham with an artificial beach of imported sand and shingle instead of building a new sea wall. On the German North Sea island of Sylt, 20 million deutschmarks were spent on beach nourishment to repair the ravages of winter storms in 1989–1990. In The Netherlands dune coasts are being protected by artificial renourishment using sand excavated from backing swales (Louisse and Van der Meulen 1991). The cost of beach renourishment will increase after the cheaper and more accessible sources of sand have been exploited, and it will be necessary to renourish beaches at shorter intervals as the sea level rise accelerates. It is likely to be too expensive to preserve the coastline where there has been a long and narrow seaside fringe of 'ribbon development' of hotels and chalets, which may have to be abandoned. It will be easier to nourish and maintain artificial beaches in coves and embayments than on long straight or gently curving sandy coasts, which will require massive breakwaters to retain sand within nourished compartments.

On the Adelaide coast the South Australian Coast Protection Board (1984) estimated that there were sufficient sand resources to maintain existing beach nourishment procedures for the first 20–30 years of the predicted sea level rise, but forecast that over a longer period attempts to renourish the beaches would be abandoned, and sea walls would have to be built to protect the coastal plain. As long as sand resources are available, however, beach renourishment may be as efficient as sea wall construction in maintaining the coastline. In The Netherlands it has been calculated that beach renourishment repeated at 5-year intervals is actually less expensive than the building and maintaining of sea walls.

Reference has been made to the extensive losses of land that will occur on the Nile delta with a sea level rise of 1 m. Existing sea walls and harbour structures will be overtopped, coast roads submerged, and built-up areas, including tourist hotels, flooded (El Sayed 1991). Beaches will disappear and the sea will inundate a wide coastal fringe. Major sea wall construction could hold the present coastline, but there will be difficulties in preserving coastal lagoons such as Lake Burrulus, which lies behind a low sandy barrier. Constructed sea walls would have to include sluices and pumping systems if the existing hydrology and ecology of this lagoon were to be artificially maintained (Sestini 1991).

As sea level rises in estuaries the bordering land will be submerged, and tides and salt water will penetrate farther upstream. These problems will only be partly overcome by building shore walls to contain a rising estuary, for this will tend to increase penetration of tides and salt water upriver. On the other hand, the building of structures to prevent sea level rising upstream, in the way that the Thames Barrier prevents storm surge flooding of London, will cause increased submergence downstream. In some estuaries navigability is likely to improve as sea level rises, providing there are not unfavourable changes in the areas of shoal deposition. In the Yangtze River, for example, maintenance of navigation upstream to Nanking will require more dredging because of increasing siltation in the wider and deeper estuary.

It will be difficult to maintain salt marshes and mangroves in their present locations as sea level rises, for this would require the artificial raising of their substrates at exactly the same rate. Any departure from this would result in changes in species proportions and zonations, and perhaps invasions by other plants. It would be possible to enclose salt marsh and mangrove areas by embankments and perpetuate their existing hydrology by pumping and draining sea water to maintain the present oscillations, but this would also be difficult and highly artificial. It may be preferable to allow salt marshes and mangroves to migrate landward on selected coastal sectors as sea level rises, displacing whatever is now in their immediate hinterlands. Alternatively, artificial intertidal shoals could be built in order to provide habitats for salt marsh and mangrove ecosystems, replacing those lost because of a rising sea level along the coastline.

On sectors of mangrove coast in southern Malaysia, where banks of earth and stone were built along the inner boundary of the mangroves to protect land developed for plantation agriculture, erosion of the mangroves has exposed the banks to wave attack. In some places they have been maintained, enlarged and armoured, and are now substantial sea walls confronting the waves of the open sea. It seems likely that efforts will be made to maintain this coastline in order to prevent the loss of these plantations.

On parts of the coast of south-east Asia where the mangrove fringe has been converted to aquaculture, maintenance of existing features as sea level rises will require the enlargement of the enclosing embankments to prevent the sea invading the fish and prawn ponds (see Fig. 82c). Pond floors will have to be raised to match the levels of the rising sea, and pumping and drainage systems introduced to control the inflow and outflow of sea water. It seems likely that the more elaborate tiger prawn ponds, managed by marine irrigation, will survive as economic units that can be maintained behind enlarged coastline walls as sea level rises, whereas the low-technology intertidal ponds will be abandoned and will disappear beneath mangroves. Maintaining the present coastline requires modifications of the ricefield zone to landward, which risks saltwater inundation as sea level rises. Further embankments can be constructed to prevent this, and freshwater irrigation of the ricefields can be maintained, but pumping will be necessary to remove water when the fields are to be drained. The attempt to maintain such a coastline will be at the expense of losing at least part of the mangrove fringe and intertidal zone (and the resources, including nearshore fisheries, associated with these) and perhaps part of the intertidal mudflat; the costs are in the elaboration of sea walls, the raising of levels in the zone of fish or prawn ponds, the building of structures to keep salt water out of the ricefield zone, and pumping systems for drainage and irrigation. It is estimated that these building costs would be of the order of $US5 million per coastline kilometre, not including the lateral bunds (running inland from the coast) that would be necessary along the boundaries of each region thus treated.

Maintaining the coastlines of low islands on coral reefs will be difficult. It may be necessary to quarry material from uninhabited cays, and limestone from some of the reefs, to provide material to build up the low islands artificially. Construction of enclosing sea walls will be expensive, because of the need to import suitable material. The readily available material, reef limestone, is permeable, and walls built of it would not prevent water rising in the enclosed land. It seems likely that only the larger, higher, and more densely populated reef islands could be protected by such structures, and maintained as increasingly artificial islands. The situation will be ameliorated, however, if sea level rise prompts upward growth of corals on surrounding reef flats, possibly after they have been planted there to assist reef revival.

On coastlines close to urban and industrial centres it is likely that a sea level rise will stimulate expenditure on structures (like the Thames Barrier) designed to prevent submergence and erosion. Dams and barrages have been built on the Dutch coast across the mouths of the distributaries of the River Rhine, and inflatable structures in the marine entrances to the Lagoon of Venice are planned to save that city from submergence and destruction by increasingly high storm tides. The devastating sea flood of 4 November 1966, which flooded St Mark's Square to a depth of 1.5 m and caused £50 million damage, initiated various responses. The canals in and around Venice, which have been shallowed by siltation, are to be deepened by dredging to reduce flooding. Barrages, known as Moses gates, are to be in place across the three entrances by 1996. These are steel canisters that will normally contain sea water and lie in a concrete trench on the channel floor, but when a storm surge is expected, compressed air will be pumped into them so that they rise to block the entrances and prevent seawater incursion to the Venice Lagoon. Pirazzoli (1991b) has predicted that as sea level rises these will become inadequate, and that the Venice Lagoon will eventually have to be protected by building a more permanent artificial barrier structure. This presupposes either the abandonment of Porto Marghera or the building of a ship canal that would split Venice Lagoon in two. The chief problem would then be maintaining water quality in a stagnant lagoon system that is already heavily polluted.

It is very likely that where land has recently been reclaimed from the sea for use by large coastal populations (as in Singapore, Hong Kong and Tokyo

Bay) strenuous efforts will be made to retain it as sea level rises by building sea walls and introducing pumping systems, using techniques familiar from the history of The Netherlands coast. The problems include the disposal of waste water, because existing disposal systems that use outfall by gravity will have to be redesigned, incorporating pumping devices, if they are to remain serviceable as sea level rises. Efforts will certainly be made to prevent the sea inundating farmland. The Yilan Plain, in north-east Taiwan, is an example of an already subsiding coastal area where dykes have been built to maintain the coastline and protect an intensively used lowland from marine invasion (Hsu 1985).

Few existing artificial structures have been designed to cope with the effects of a 1-m sea level rise. At Galveston, for example, the existing oceanside sea wall will be overtopped and outflanked by larger storm surges after a 1-m sea level rise, and extensive supplementary works will be necessary to hold the coastline (Leatherman 1984), as well as further landfill to raise the ground surface. Gibbs (1986) examined the planning issues arising from attempts to prevent the submergence of Charleston, South Carolina, a coastal city which needs to be protected from the sea, but also a port which will require continuing sea access. Where ports are to be maintained it will be necessary to raise quays and piers to match sea level rise so that shipping operations can continue.

Sea walls have also been built to maintain the coastline where coastal subsidence has resulted in recurrent flooding, with pumping schemes to dispose of surplus water, as near Bangkok in Thailand (Ramnarong and Buapeng 1991), Taipeh in Taiwan, Tokyo and Niigata in Japan, and the deltaic north coast of Java, east of Jakarta. Where such structures have already been built, it is likely that efforts will be made to maintain them against a rising sea level. There will be severe problems in maintaining towns that are already below high tide level, such as Georgetown in Guyana, Port-au-Prince in Haiti and Puerto Cortes in the Honduras, with elaborate sea defences.

If maintaining the coastline becomes a widespread endeavour, the prospect is one of continuing elaboration of sea walls, a sequence familiar from the history of The Netherlands coast, where coastal defences have become successively higher and wider to counter a continuing relative sea level rise (Fig. 66, page 112). It has been estimated that the raising and elaboration of coastal defences to match a sea level rise of 20 cm along about 250 km of coastline would cost about $US1 billion: for a 1-m rise the cost would be about $US10 million per kilometre (Goemans 1986).

Scenario 3: Counter-attack

The cost of maintaining the coastline by building and elaborating sea walls and putting in drainage and pumping systems to prevent marine submergence as sea level rises will be so great that it is difficult to envisage many coastal countries achieving it on a large scale without substantial international assistance. An alternative may be to counter-attack by constructing sea walls offshore, and reclaiming the enclosed shallow areas for productive use. Where this is possible, the economic returns from the land gained could offset the costs of building sea walls and associated structures. A well-known example is the enclosure of the marine embayment formerly known as the Zuider Zee in The Netherlands by a dam, completed in 1932, to form freshwater Lake Ijssel, the shores of which have been progressively advanced by dykes to contain areas of reedswamp, in due course reclaimed as polders for agriculture.

Land reclamation schemes have already been extensive on some coasts, notably at the head of Tokyo Bay and on the southern and eastern shores of Singapore Island (Wong Poh Poh 1985). At the Commonwealth Heads of Governments Conference in 1989 the Malaysian Prime Minister Mahathir Mohamad suggested that a series of sea walls about 3.2 km (later revised to 2.2 km) offshore would be constructed along the Strait of Melaka coast of Peninsular Malaysia over the next 30 years as a prelude to large-scale reclamation of the shallow nearshore area there. It would be necessary to enclose compartments between major river estuaries, and there would be losses of existing intertidal and nearshore ecosystems, notably mangroves and mudflats, with their associated fish and shellfish resources, but there would be substantial economic returns from the reclaimed land, used for agriculture, aquaculture, and urban and industrial development. The enclosing walls would have to be built to levels that excluded a rising sea.

A similar approach could be applied on the coast of the Bight of Bangkok, where an offshore sea wall could permit large-scale land reclamation. It would be necessary to design and build a new drainage and irrigation system for the Bangkok region, adjusted to the rise of sea level (Fig. 84) (Bird 1989c).

Other areas where a counter-attack strategy may prove attractive include Jakarta Bay in Indonesia and shallow areas in Hong Kong. It has been suggested that a sea level rise might prompt the enclosure and reclamation of Port Phillip Bay in Australia, which has a 260-km coastline and a marine entrance only 3 km wide (Fig. 85): it would certainly be much less expensive to defend the narrow entrance against a rising sea than to maintain the existing lengthy coastline. The broad tidal mudflats that front the city of Cairns, in north-eastern Australia, may also tempt land reclamation as a counter-attack to sea level rise (Fig. 86).

Another response has been to raise land levels to offset the effects of submergence. Reference has been made to 'raising the grade' in Galveston, Texas. There has been a proposal to raise the level of Poveglia Island in the Venice Lagoon by injecting concrete into the ground, and similar techniques have been used to raise a Norwegian oil rig in the North Sea: it was elevated 6 m by 200 jacks in a few days at a cost of $US500 million. The Temple at

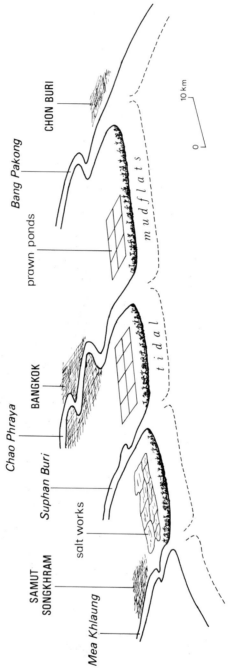

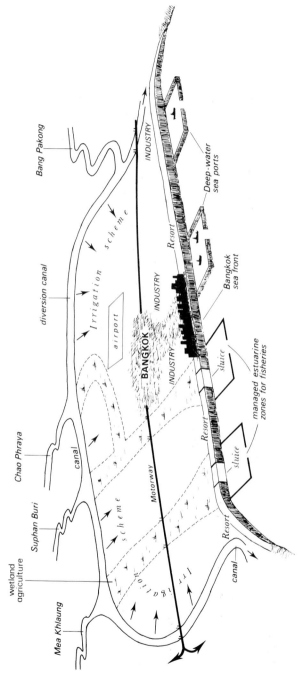

Figure 84   The coastal region at the head of the Gulf of Thailand as it is now (1990), and as it would be a century hence if the response to sea level rise is to 'counter-attack' by building a sea wall offshore and reclaiming the tidal mudflats

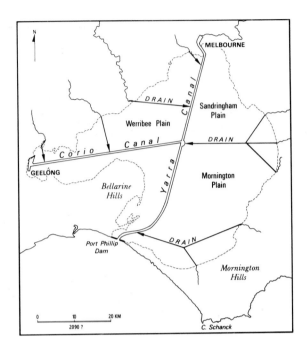

Figure 85   The 'Dutch solution' in Port Phillip Bay, Australia. As sea level rises it may become so difficult and expensive to maintain the existing 256-km bay coastline that a dam will be built across the 3-km entrance from the sea. In such circumstances, canals could be built to maintain ship access to the ports of Melbourne and Geelong (much like the waterways that lead to the Dutch ports of Amsterdam and Rotterdam), and drains to carry water from inflowing streams. The intervening areas could then be reclaimed as polders. This suggestion is highly unpopular in Melbourne

Abu Simbel has been artificially raised to prevent it being damaged by the rising waters of Lake Nasser, a Nile reservoir, at a cost of $US50 million.

Raising ground levels requires a supply of suitable sediment, rock debris or rubble waste. Sand can be pumped in from sea floor areas, using the same techniques as for beach nourishment, as in Singapore, or brought from hinterland quarries, as in Hong Kong. Material for coastal land reclamation in Singapore has also been imported from nearby Indonesian islands and from Malaysia.

On the coasts of south-east Asia where extensive fish and prawn ponds have been constructed, counter-attack by building an offshore sea wall, at or below the existing low tide line, envisages the reclamation of the intertidal mudflats and the mangrove fringe for productive purposes, usually additional fish or prawn ponds (see Fig. 82d). These would either have to be served with sea irrigation and drainage pumping systems, or artificially raised to a level appropriate for

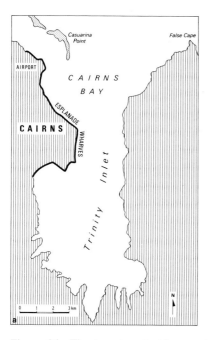
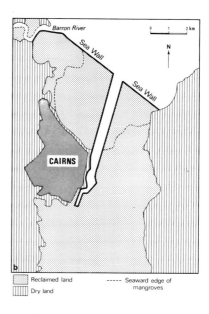

Figure 86   The 'counter-attack' scenario applied to the coast of Cairns, north-east Australia. At present Cairns has an esplanade facing across the shallow waters of Cairns Bay, where extensive mudflats are exposed at low tide. Sea level rise would submerge mudflats and mangrove areas (a), but could prompt the construction of sea walls and the reclamation of tidal mudflats, while maintaining a canal to serve the port (b)

gravitational inflow at high tide and outflow at low tide. The intertidal zone and the mangroves, together with the resources they generate at present, will be lost, but increased production from the reclaimed area could in due course at least partly offset these losses and the costs of building the sea wall and reclaiming the intertidal area. The initial outlay in building costs would be of the order of $US7 million per coastline kilometre (plus the cost of a lateral boundary wall), but this could be treated as an income-generating investment rather than a simple capital outlay.

There will be objections to schemes that require large-scale reclamation of intertidal areas and losses of salt marsh and mangroves, bearing in mind that these ecosystems sustain fisheries and shellfish resources, that they yield a variety of useful products, and that they are of importance for nature conservation. In practice, it may be possible to design a compromise, in which some sectors of coastline are reclaimed while other, intervening, sectors are kept as nature reserves within which salt marshes, mangroves and intertidal zones can migrate landward as the sea rises (Fig. 87).

1990

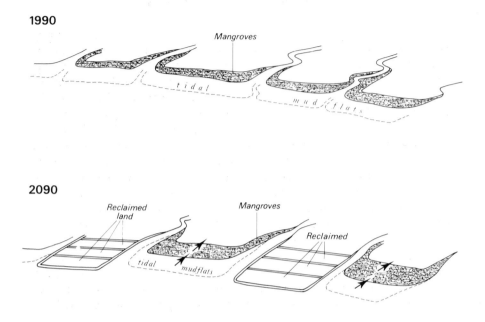

2090

Figure 87   Modification of the proposal put forward in Fig. 84 to leave sectors of natural coastline where a rising sea will drive the mangrove and mudflat zones landward, and cut into beach-ridge terrain. Such a scheme would allow for the perpetuation of natural ecosystems between reclaimed coastal segments, both for nature conservation and for the sustenance of fish and shellfish resources

On coral reefs, a counter-attack strategy for low islands, such as the Maldives, could be to enclose and build up artificial land areas on a larger proportion of the reef foundation, particularly on patch reefs where the existing natural cay takes up only a small part of the reef flat area. It may be possible to fill in small atoll lagoons to form a solid artificial island structure. Management of reef areas to facilitate such a positive response to sea level rise should include protection from damaging impacts, pollution control, and the planting of corals to assist their revival on reef flats. Apart from ecological considerations, coral reefs are important sinks for carbon dioxide, extracted from the atmosphere by way of marine carbonates, and an increase in living coral reefs could help to diminish greenhouse gases and global warming.

Another form of counter-attack designed to offset the effects of sea level rise could be the construction of artificial submerged reefs offshore to maintain an array of habitats for the various marine species that may be threatened by the fact that the sea is deepening.

SUMMARY AND CONCLUSIONS

The human responses that have been discussed will be complicated by many and varied socio-economic and political considerations, and will especially be influenced by the experience gained during the early stages of a sea level rise, say 10–20 cm early in the 21st century (Jones 1991). In terms of the first scenario, evacuate and adapt, there will be problems arising from the displacement of coastal people and their migration inland to higher areas along the coast, or to other (especially urban) areas, and competition with people already living in the hinterland or other resettled areas for living space, food and energy, and employment. Some problems are already evident where people have been displaced from the coastal fringe by erosion, or by submergence due to land subsidence. There is competition for land and living space, especially in the immediate hinterland, where existing occupants face problems as their coastal neighbours start to move back inland. People will also be disadvantaged by changes accompanying a sea level rise, such as the upward movement of groundwater and increasing salinity landward of the receding coastline. In economic terms there are the costs of land losses through submergence and erosion, of evacuation and resettlement of people displaced, and of changes in land use, including the intensification of production from a reduced hinterland.

In the second scenario, maintain the coastline, these losses do not occur, but there are the costs of building and maintaining structures to prevent submergence and erosion, of instituting pumping and drainage systems, or land raising and beach renourishment projects; all at least partly offset by the benefits of continued production and occupance of the maintained coastal fringe. There are, of course, other changes that will occur in maintained coastal areas that are not directly related to the effects of sea level rise. Some will be the outcome of climatic changes resulting from the greenhouse effect, but many will be responses to other changes, whether environmental or social.

The third scenario depends on the large costs of sea wall construction and the reclamation and maintenance of reclaimed land (e.g. pumping and drainage systems) being at least partly met (or even exceeded) by the benefits of new land for production and occupation. Again the outcome will be influenced by other environmental and social changes unrelated to sea level rise, but countries that have the opportunity to counter-attack by reclamation of intertidal and nearshore areas, and have the resources to carry out such work, will be tempted to seek the territorial and economic gains that would ensue. As has been noted, there are environmental and ecological issues that need to be considered before large-scale coastal reclamation proceeds, but many of these could be met by a compromise that allows some sectors to persist as nature reserves and managed ecosystems while the coast as a whole becomes more artificial.

The human response to sea level rise will be complicated by a number of other factors, which can be mentioned only briefly here. The first of these is geographical. Even on sectors of coastline that are essentially similar, there will be regional variations in the human response to sea level rise depending on existing population density, intensity of use and extent of development. Some activities will change because they become uneconomic, or because of social and political factors: seaside resorts, for example, may become less fashionable.

The second is political. Some countries will be severely affected, even by quite a small sea level rise, and will wish to react quickly to maintain their resources (Yohe 1990). These include Bangladesh, the low-lying ocean island states such as the Maldives and Kiribati, and those with extensive deltas and coastal lowlands, such as Egypt, Indonesia, Malaysia and The Netherlands. Of these, only The Netherlands has the experience, technology and resources to maintain its coastline and preserve its land area in response to a continuing sea level rise (Goemans 1986). Elsewhere, people may be driven inland to other regions, or overseas. Countries with steep and high coasts, and only limited lowlands, such as Japan, the Philippines and New Zealand, will be less severely affected, although the problems will still be great on their densely-populated lowland sectors. Countries with no coastline, such as Switzerland and Bolivia, cannot be expected to take much notice of these problems. Differences in political response became evident during the discussions of the group of 77 states considering restrictions on carbon dioxide emissions at the United Nations Conference on Environment and Development (UNCED) in Rio in June 1992.

There are also sharp contrasts in human response to sea level rise in terms of economic situations. There will inevitably be national differences in the ways in which people and governments respond to the problems of a rising sea level (Vellinga 1988). In the United States and other western countries the problems will be seen mainly in terms of insurance costs (e.g. through an evolution of the American National Flood Insurance Program), legislation, compensation, and planning issues. In Third World countries, where basic requirements of health and food security loom larger and the problems of displaced and impoverished people will be severe, it is likely that assistance will be sought from the international community, bearing in mind that sea level rise as a consequence of the greenhouse effect is primarily a by-product of industrial pollution generated by western countries.

There are variations related to the existing status of planning and coastal management in various countries. Over the next few decades the best response to the predicted sea level rise may be to start reorganising coastal land use planning in low-lying coastal areas by drawing set-back lines determined in relation to expected submergence and erosion. For example, it is unwise to develop new seaside resorts within 200 m of the present high tide coastline on beach-ridge terrain unless they are designed to be abandoned or relocated as submergence and erosion proceed. Large and expensive hotels designed to attract luxury tourism should be built on high ground, rather than on low-lying beach-ridge terrain. Attention to planning and coastal management issues increases sharply when events such as the 1953 North Sea storm surge occur, and is likely to increase when the reality of a global sea level rise becomes obvious, especially where hazardous events such as storm surges become increasingly frequent and severe (Eid and Hulsbergen 1991). There seems to be a preference for engineering solutions to the problems of a rising sea level. This generally means the building and elaboration of sea walls, and these are likely to become more extensive, at least on some parts of the coast. Coastlines will thus become increasingly artificial. While there has been much talk of 'soft engineering', such as artificial beach renourishment, revegetation of dunes, and dumping of dredged sediment to form coastal wetlands, the emphasis remains on solid structures to counter erosion and submergence.

Environmental planning is necessary if natural, scenic and cultural features are to be conserved, and management is required to maintain areas of salt marsh, mangrove and intertidal mudflats and sandflats, to preserve sites for nature conservation, and to ensure the continued existence of beaches and coastal dunes. There is already a marked trend toward the artificialisation of coastlines, and sea level rise is likely to accelerate this trend. Yet most structures so far built in coastal areas threatened by submergence and erosion were not designed to last for more than a few years or decades, and their abandonment in the face of the rising sea could prove much less expensive than the building of large sea walls, and accompanying pumping and drainage systems. Beach management and renourishment will be economic only in areas of intensive recreational and resort development, while beaches that at present provide resources for local people and small numbers of visitors could be allowed to disappear.

On the global scale, the most widespread response to sea level rise is likely to be gradual evacuation and abandonment, with accompanying social adaptation and land use changes. However, where there is already intense population pressure, in many low-lying coastal regions, an unwillingness to surrender large areas of coastal lowland to an encroaching sea may prompt demands for the construction of sea walls along selected parts of the coast, or even offshore, as a counter to sea level rise.

Human responses to sea level rise will also be affected by technical changes and innovations over the next few decades. The problems and costs of submergence and erosion by a rising sea will stimulate much reorganisation of activities in coastal areas. An example is where low-lying coastal land has been developed for primitive, labour-intensive fish or prawn ponds using high tide irrigation and low tide drainage. At present such activities contribute directly to the food supply, as well as to the income, of local people. As sea level rises, the cost and effort of maintaining such ponds are likely to favour selective conversion to more complex, localised, highly mechanised aquaculture, developed and operated by companies employing relatively few people and generating an expensive product for sale elsewhere.

Some of the pressure for development and utilisation of coastal areas may be reduced as a result of changing attitudes to beach and coastal recreation. For example, the publicity about increasing ultraviolet radiation as a result of the springtime thinning of the ozone layer in high southern latitudes, and the consequent increase in the risk of skin cancer, has already modified beach behaviour in Australia, where the acquisition of a suntan was formerly a greater incentive to beach recreation than it is now. Although no statistics are available, it has been suggested by the Australian media that the numbers of people visiting, and staying on, beaches has diminished during the last few summers (Bird 1988d). Increasing thermal discomfort may lead to some abatement in the demands for tourist facilities around the Mediterranean and the Caribbean, which would reduce problems of pollution and excessive coastal urbanisation, but also lead to a decline in economic returns to countries that have benefited from the tourist boom in recent decades.

Sea level rise may stimulate innovations, such as the breeding of salt-tolerant crops cultivable in sea water or saline soils, new products from mangrove agroforestry, and the generation of organic materials from algal cultures in marine enclosures. Submergence is likely to create more extensive nearshore shallow environments, which could be put to productive use for fish and shellfish cultivation. It seems likely that there will be new human settlements afloat, or on artificial islands.

The human response to sea level rise will also be influenced by changing local, regional and global economic circumstances, some of which seem readily predictable (such as continuing growth of the human population), but many quite unforeseeable. If there is indeed an increase in world population from the present 7 billion to more than 20 billion during the next few decades there will be enormous environmental problems, and accompanying socio-economic and political changes that are very difficult to forecast. Even on the coast, the human responses to changes that would have occurred if existing environments persisted will be modified by the effects of a rising sea level. What actually occurs will be a response to overall circumstances, of which coastal changes due to sea level rise will be only one component. Nevertheless, there are some distinctive issues in coastal regions that will have to be addressed if sea level rise occurs. Some require a physical response, such as the building of sea walls and drainage systems; others a socio-economic response, such as the movement of people, products, and money: and all will require the framing and application of policy by decision makers.

Global economic circumstances will also affect human responses to sea level rise, with variations between coastal nations (Gibbs 1984). There will be changes in production, demand, and trade of goods and services, which will have effects on coastal development. Human migrations, induced by political as well as economic factors, have already led to movements of people towards coastal areas, notably in Chile, Somalia and China. If increased urbanisation and industrial development occur on and near the coast, the population will rise, and there will be an intensifying demand that the coastline be maintained, or advanced by reclamation schemes, rather than allowed to retreat as the sea rises. Fluctuating demands for coastal and marine products such as salt, fish and shellfish, mangrove timber, sand, gravel, limestone and various minerals, will have socio-economic consequences for coastal inhabitants, and implications for coastal management. Changes in trading conditions and in the means whereby commodities are traded could lead to changes in the development of ports and harbours, together with their associated urban areas. Jones (1991) offered a framework for an economic approach to the impacts of sea level rise (Fig. 88).

Many coastlines are national frontiers, and changes due to a sea level rise are likely to influence some national maritime claims, especially on coasts where the nearshore area is very shallow (Bird and Prescott 1989). Historically, coastal areas have been influenced by invasions, illegal immigration and piracy, and the fortification of coasts has been a recurrent need. The issues that have been discussed in terms of sea level rise effects could be modified if political

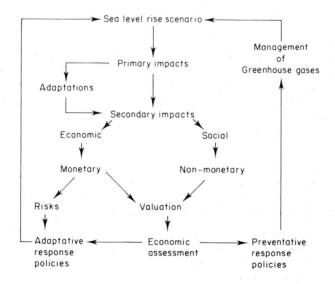

Figure 88   A framework for an economic approach to the impacts of sea level rise, based on Jones (1991)

circumstances change the strategic significance of a nation's coastline, and this could have implications for coastal management. In the event of war, or the threat of war, a coastline that has been worth maintaining by structural works is likely to be worth defending by fortifications and anti-invasion structures.

Broadus (1988) commented on the choices available to decision makers confronted with problems of a forecast sea level rise. Some will seek to make gradual or incremental adjustments to sea level rise as it develops, allocating resources to adapt to it (e.g. by constructing sea walls) as time goes by. Others will say that it is necessary to seek more accurate information (e.g. from further scientific research, surveys and monitoring), meanwhile investing in insurance and reducing commitments by restricting further coastal development. A third group will concern themselves with the problems arising from evacuation of submerging and eroding coastal areas. A fourth group will focus on the need to defend and maintain the coastline, or even advance it seawards, and urge investment in structures that will prevent erosion and submergence. They may be opposed by those who would like to avoid using present resources to build such structures, and let our descendants deal with the problem. There will also be those who feel the problems can be left to organisations that have been established to deal with coastal planning and management.

Overall, it appears that losses will exceed gains as sea level rise increases erosion and submerges coastal areas. There may be some ecological gains from changes in coastal and marine ecosystems, and economic gains where harbours are deepened; and certain groups such as coastal engineers and industries concerned with sea wall construction, beach nourishment, land reclamation, and pumping and drainage systems will obviously benefit. The losses of coastal land, the impoverishment of many coastal and marine ecosystems, the costs arising from erosion and submergence of developed coasts, the costs incurred in evacuating and resettling coastal people, and the costs of coastal maintenance and reclamation will all become very considerable as sea level rises.

As awareness of the problems increases there is a growing interest in ways in which the problem could be solved, or at least reduced, by curbing its cause, the enhanced greenhouse effect. There are demands for the reduction of emission of gases such as carbon dioxide, methane, nitrous oxide and the chlorofluorocarbons from industrial, domestic and agricultural operations, and international discussion of the actions and agreements required to achieve this featured strongly at the UNCED conference in Brazil in 1992. Clearing of forests, especially in the tropics, has been seen as a cause of increased emissions of greenhouse gases, and an improved global vegetation cover and intensified coral growth could help by extracting more carbon dioxide from the atmosphere. It has been suggested that more water should be retained in land reservoirs and areas below sea level such as the Dead Sea and Lake Eyre, but this would delay only slightly the global sea level rise. Highly technological solutions, such as the pumping of sea water into the very cold centre of Antarctica, where it would be retained as ice, may be appealing, but are usually vastly expensive. More practical is the possibility of switching to forms of energy use that do not generate greenhouse gases, notably solar, wind, wave and tidal energy, or the greater use of nuclear energy. Increased economies in the use of fossil fuels are widely advocated, and alternative fuels such as hydrogen, alcohol and other recyclable organic compounds are under consideration. The difficulties of international co-operation on the control of the greenhouse effect are formidable, but they are really only a first step towards integrating the global management of the land, the air and the oceans.

Meanwhile, it seems likely that the enhanced greenhouse effect will develop over the next few decades, and that global sea level will rise. This study has examined the possible consequences around the world's coastlines, and the possible human responses. There are many uncertainties, and predictions can be made only in general terms, with nested scenarios. As sea level rise proceeds, the changes that occur around the world's coastlines will be monitored, and models of geomorphological and ecological response progressively refined. Much more research is needed on human responses to the geomorphological and ecological changes that will occur on each type of coastline as sea level rises, but it is already clear that careful planning and management of coastal areas could reduce the adverse socio-economic and political impacts. There is a need to document the ways in which the human responses are modified around the world's coastline by accompanying social, economic and political changes, some in response to environmental modifications (other than sea level rise) caused by global warming, and others that would have taken place quite independently (Wyrtki 1990). The forecasting of these wider social, economic and political changes is beyond the scope of this study, but it is noted that such changes are likely to take place more rapidly than the predicted sea level rise by 1 m over the coming 100–150 years.

In 1991 the Intergovernmental Panel on Climate Change commissioned a Response Strategies Working Group, which produced the ten recommendations for future coastal and ocean management listed in Table 2. It is for the present generation to begin to assess the problems and institute the necessary monitoring and modelling. The solutions, within a framework of international environmental management, will be found and implemented in the changing conditions of the 21st century, of which climate modification and sea level rise will be components. At this stage the debate shows encouraging signs of global awareness: the prospect of submerging coasts is already widely acknowledged, and it is hoped that the present study will contribute to further discussion, research and management studies around the world's changing coastlines.

Table 2   Recommendations of IPCC Response Strategies Working Group (1991)

---

1. By the year 2000 coastal nations should implement coastal management plans.
2. High risk coastal areas should be mapped and assessed.
3. Future sea level rise should be factored into plans for coastal development to reduce future vulnerability.
4. Plans for emergency preparedness for coastal storms should be strengthened and include future climate change.
5. An international focus on the effects of sea level rise should be augmented.
6. Technical assistance should be provided to coastal nations in developing plans to deal with sea level rise and other effects of climate change.
7. International co-operation and education is needed to limit population growth and development in coastal regions.
8. Research on understanding and predicting the impacts of climate change on sea level rise should be strengthened.
9. A global ocean and coastline monitoring network should be implemented.
10. Data and information on climate change and sea level rise should be widely disseminated and available for preparation of coastal management plans.

References

Aksornkoae, S. (1988) Mangrove habitat degradation and removal in Phangna and Ban Don Bays, Thailand. *Tropical Coastal Area Management*, **3**(1): 16.

Armentano, T.V., Park, R.A. and Cloonan, C.L. (1986) *The effects of future sea level rise on US coastal wetland areas*. Holcombe Research Institute, Indianapolis, USA.

Baker, G. (1943) Features of a Victorian limestone coastline. *Journal of Geology*, **51**: 359–386.

Barnett, T.P. (1984) The estimation of 'global' sea level changes: a problem of uniqueness. *Journal of Geophysical Research*, **89**: 7980–7988.

Barth, M.G. and Titus, J.G. (1984) *Greenhouse Effect and Sea Level Rise*. Van Nostrand Reinhold, New York.

Bird, E.C.F. (1962) The utilisation of some Australian coastal lakes. *Australian Geographer*, **8**: 199–206.

Bird, E.C.F. (1978) *The Geomorphology of the Gippsland Lakes*. Ministry for Conservation, Victoria, Australia.

Bird, E.C.F. (1982) Foundations of the Nepean Peninsula. In: C.N. Hollinshed (ed.) *Lime, Land, Leisure*. Flinders, Victoria, Australia, pp. 1–24.

Bird, E.C.F. (1984) *Coasts*. Blackwell, Oxford.

Bird, E.C.F. (1985) *Coastline Changes*. Wiley–Interscience, Chichester.

Bird, E.C.F. (1986) Potential effects of sea level rise of the coasts of Australia, Africa and Asia. In: J.G. Titus (ed.) *Effects of Changes in Stratospheric Ozone and Global Climate*, vol. 4. US Environment Protection Agency, Washington, DC, pp. 83–98.

Bird, E.C.F. (1987) The world's disappearing beaches. *1987 Yearbook of Science and the Future*. Encyclopaedia Britannica, Chicago, pp. 114–127.

Bird, E.C.F. (1988a) The tubeworm *Galeolaria caespitosa* as an indicator of sea level rise. *Victorian Naturalist*, **105**: 98–104.

Bird, E.C.F. (1988b) Physiographic indications of a sea-level rise. In: G.I Pearman (ed.) *Greenhouse: Planning for Climatic Change*. CSIRO Division of Atmospheric Research, Melbourne, Australia, pp. 60–73.

Bird, E.C.F. (1988c) Effects of a sea level rise in Port Phillip Bay. In: *Making the Most of the Bay*. Ministry for Planning and Environment, Victoria, Australia.

Bird, E.C.F. (1988d) Man's response to changes in the Australian coastal zone. In: K. Ruddle, W.B. Morgan and J.R. Pfafflin (eds.) *The Coastal Zone: Man's Response to Coastal Change*. Harwood Academic Publishers, New York, pp. 365–382.

Bird, E.C.F. (1989a) Coastal geomorphology in the humid tropics. In: E.C.F. Bird and D. Kelletat (eds.) *Zonality of Coastal Geomorphology and Ecology*. Essener Geographischen Arbeiten, vol. 18, pp. 31–58.

Bird, E.C.F. (1989b) The effects of sea level rise on the Estonian coastline. *Proceedings of the Estonian Academy of Sciences*, **38**: 141–149.

Bird, E.C.F. (1989c) The effects of a rising sea level on the coasts of Thailand. *ASEAN Journal on Science and Technology for Development*, **6**: 1–13.

Bird, E.C.F. (1992) The impacts of sea level rise on coral reefs and reef islands. In: P. Fabbri (ed.) *Ocean Management in Global Change*. Elsevier Applied Science, Amsterdam, pp. 90–107.

Bird, E.C.F. and Barson, M.M. (1975) Shoreline changes in Westernport Bay. *Proceedings of the Royal Society of Victoria*, **87**: 15–28.

Bird, E.C.F. and Guilcher, A. (1982) Observations préliminaires sur les récifs frangeants actuels du Kenya et sur les formes littorales associées. *Revue de Géomorphologie Dynamique*, **31**: 113–135.

Bird, E.C.F. and Koike, K. (1986) Man's impact on sea level changes: a review. *Journal of Coastal Research*, Special Issue **1**: 83–88.

Bird, E.C.F. and Missen, G.J. (1990) The effects of a predicted sea level rise on the coasts of south-east Asia: socio-economic and policy considerations. In: Chou Loke Ming (ed.) *Task Team Report on Climatic Studies in the East Asian Seas*. UNEP, pp. 213–232.

Bird, E.C.F. and Ongkosongo, O.S.R. (1980) *Environmental changes on the coasts of Indonesia*. United Nations University.

Bird, E.C.F. and Prescott, J.R.V. (1989) Rising global sea levels and national maritime claims. *Marine Policy Reports*, **1**(3): 177–196.

Bird, E.C.F. and Rosengren, N.J. (1987) Coastal cliff management: an example from Black Rock Point, Melbourne, Australia. *Journal of Shoreline Management*, **3**: 39–51.

Bird, E.C.F. and Schwartz, M.L. (eds.) (1985) *The World's Coastline*. Van Nostrand Reinhold, New York.

Bloom, A.L. (1979) *Atlas of Sea Level Curves*. Cornell University, New York.

Boorman, L.A., Goss-Custard, J.D. and McGrorty, S. (1989) *Climatic change, rising sea level and the British coast*. Research Publication no. 1, Institute of Terrestrial Ecology, HMSO, London.

Broadus, J. (1988) Economizing human responses to subsidence and rising relative sea level. In: J.D. Milliman (ed.) *Sea-Level Rise and Coastal Subsidence: Problems and Strategies*. Papers presented to the SCOPE Workshop on Subsiding Coasts, Bangkok, November 1988 (unpublished).

Broadus, J., Milliman, J., Edwards, S., Aubrey, D. and Gable, F. (1986) Rising sea level and damming of rivers: possible effects in Egypt and Bangladesh. In: J.G. Titus (ed.) *Effects of Changes in Stratospheric Ozone and Global Climate*, vol. 4. US Environment Protection Agency, Washington, DC, pp. 165–189.

Brooke, J.S. (1991) *Coastal Defence: The Retreat Option*. Paper presented to the Institution of Water and Environmental Management, winter meeting 1991, Reading, UK.

Brown, B.E. (ed.) (1990) Coral bleaching. *Coral Reefs*, **8**: 153–232.

Brunsden, D. and Jones, D.K.C. (1980) Relative time scales and formative events in coastal landslide systems. *Zeitschrift für Geomorphologie, Supplementband*, **34**: 1–19.

Bruun, P. (1962) Sea level rise as a cause of shore erosion. *Proceedings of the American Society of Civil Engineers (Waterways and Harbors Division)*, **88**: 117–130.

Bruun, P. (1988) The Bruun Rule of erosion by sea level rise: a discussion of large-scale two- and three-dimensional usages. *Journal of Coastal Research*, **4**: 627–648.

Bryant, E.A. (1983) Regional sea level, Southern Oscillation and beach changes, New South Wales, Australia. *Nature*, **305**: 213–216.

Bryant, E.A. (1985) Rainfall and beach erosion relationships, Stanwell Park, Australia, 1895–1980: worldwide implications for coastal erosion. *Zeitschrift für Geomorphologie, Supplementband*, **57**: 51–65.

Bryant, E.A. (1988) Sea surface temperature and high-tide beach changes, Stanwell Park, Australia, 1943–1978. *Journal of Coastal Research*, **4**: 229–243.

Budd, W.F. (1988) The expected sea-level rise from climatic warming in the Antarctic. In: G.I Pearman (ed.) *Greenhouse: Planning for Climatic Change*. CSIRO Division of Atmospheric Research, Melbourne, Australia, pp. 74–82.

Budd, W.F. (1991) Antarctica and global change. *Climatic Change*, **18**: 271– 299.

Buddemeier, R.W. and Hopley, D. (1989) Turn-ons and turn-offs: causes and mechanisms of the initiation and termination of reef growth. In: *Proceedings of the 6th International Coral Reefs Symposium*, vol. 1, pp. 253–262.

Buddemeier, R.W. and Smith, S.V. (1988) Coral reef growth in an era of rapidly rising sea level: predictions and suggestions for long-term research. *Coral Reefs*, **7**: 51–56.

Burton, T. (1982) Mangrove changes recorded north of Adelaide. *Safic*, **6**: 8–12.

Cambers, G. (1976) Temporal scales in coastal erosion systems. *Transactions, Institute of British Geographers*, **1**: 246–256.

Carter, C.H. and Guy, D.E. (1988) Coastal erosion: processes, timing and magnitude at the bluff toe. *Marine Geology*, **84**: 1–17.

Carter, R.W.G. (1988) *Coastal Environments*. Academic Press, New York.

Carter, R.W.G. (1991) Near-future sea-level impacts on coastal dune landscapes. *Landscape Ecology*, **6**: 9–40.

Carter, R.W.G. and Devoy, R.J.N. (eds.) (1987) The hydrodynamic and sedimentological consequences of sea level rise. *Progress in Oceanography*, **18**: 136–154.

Cartwright, D.E. (1974) Years of peak astronomical tides. *Nature*, **248**: 656–657.

Cavazzoni, S. (1983) Recent erosive processes in the Venetian lagoon. In: E.C.F. Bird and P. Fabbri (eds.) *Coastal Problems of the Mediterranean Sea*. University of Bologna, pp. 19–22.

Cencini, C., Cuccoli, L., Fabbri, P., Monari, F., Sembeloni, F. and Torresani, S. (1979) *Le spiagge della Romagna, uno spazio da proteggere*. Consiglio Nat. Ricerche, Bologna.

Chappell, J. (1974) Geology of coral terraces, Huon Peninsula, New Guinea: a study of Quaternary tectonic movements and sea level changes. *Bulletin of the Geological Society of America*, **85**: 553–570.

Chappell, J. (1982) Evidence for a smoothly falling sea-level relative to north Queensland, Australia, during the past 6000 years. *Nature*, **302**: 406–408.

Chappell, J. and Veeh H.H. (1978) Late Quaternary tectonic movements and sea level changes at Timor and Atauro Island. *Bulletin of the Geological Society of America*, **89**: 356–368.

Chen Jiyu, Lu Cangzi and Yu Zhiying (1985) China. In: E.C.F. Bird and M.L. Schwartz (eds.) *The World's Coastline*. Van Nostrand Reinhold, New York.

Chin Kuo (1986) Flooding in Taipeh, Taiwan and coastal drainage. In: J.G. Titus (ed.) *Effects of Changes in Stratospheric Ozone and Global Climate*, vol. 4. US Environment Protection Agency, Washington, DC, pp. 37–46.

Clayton, K.M. (1989) Implications of climatic change in coastal management. Paper presented to Institute of Civil Engineers Conference, Bournemouth, May 1989.

Cocks, K.D., Gilmour, A.J. and Wood, N.H. (1988) Regional impacts of rising sea levels in coastal Australia. In: G.I Pearman (ed.) *Greenhouse: Planning for Climatic Change*. CSIRO Division of Atmospheric Research, Melbourne, Australia, pp. 105–120.

Constanza, R., Sklar, F.H. and White, M.L. (1990) Modelling coastal landscape dynamics. *Bioscience*, **40**: 91–107.

Cubit, J.D. (1985) Possible effects of recent changes in sea level on the biota of a Caribbean reef flat, and predicted effects of rising sea levels. In: *Proceedings of the 5th International Coral Reef Congress*, vol. 3, pp. 111–118.

Daly, R.A. (1934) *The Changing World of the Ice Age*. Yale University Press, New Haven.

Dana, J.D. (1872) *Corals and Coral Islands*. Dodd Mead and Co., New York.

Daniels, R.C. (1992) Sea-level rise on the South Carolina coast: two case studies for 2100. *Journal of Coastal Research*, **8**(1): 56–70.

Darmody, R.G. and Foss, J.E. (1979) Soil–landscape relationships of the tidal marshlands of Maryland. *Journal of the Soil Science Society of America*, **43**: 534–541.

Darwin, C. (1842) *The Structure and Distribution of Coral Reefs*. Smith, Elder and Co., London.

Davies, J.L. (1972) *Geographical Variation in Coastal Development*. Oliver and Boyd, Edinburgh.

Davies, P.J. (1983) Reef growth. In: D.J. Barnes (ed.) *Perspectives on Coral Reefs*. Manuka, Clouston, pp. 69–106.

Davis, D.W., McCloy, J.M. and Craig, A.K. (1988) Man's response to coastal change in the northern Gulf of Mexico. In: K. Ruddle, W.B. Morgan and J.R. Pfaffin (eds.) *The Coastal Zone: Man's Response to Change*. Harwood Academic Publishers, New York.

Davis, W.M. (1928) *The Coral Reef Problem*. Special Publication no. 9, American Geographical Society.

Day, J.W. and Templet, P.H. (1990) Consequences of sea level rise: implications for the Mississippi delta. *Developments in Hydrobiology*, **57**: 155–165.

Dean, R.G. and Maurmeyer, E.M. (1983) Models for beach profile response. In: P.D. Komar (ed.) *Handbook of Coastal Processes and Erosion*. CRC Press, Boca Raton, Florida, pp. 151–165.

Delft Hydraulics Institute (1991) *Implications of Relative Sea Level Rise on the Development of the Lower Nile Delta, Egypt*. Report H927.

Department of the Environment, UK (1991) *The Potential Effects of Climatic Change in the United Kingdom*. HMSO, London.

De Ronde, J.G. (1988) Past and future sea level rise in the Netherlands. Paper presented to the Working Group on Sea Level Rise and Coastal Processes, Palm Coast, Florida, March 1988.

De Ronde, J.G. (1991) *Rising Waters: Impacts of the Greenhouse Effect for the Netherlands*. Rijkswaterstaat, The Hague, The Netherlands.

De Sylva, D. (1986) Increased storms and estuarine salinity and other ecological impacts of the greenhouse effect. In: J.G. Titus (ed.) *Effects of Changes in Stratospheric Ozone and Global Climate*, vol. 4. US Environment Protection Agency, Washington, DC, pp. 153–164.

Donn, W.L., Farrand, W.R. and Ewing J.R. (1962) Pleistocene ice volumes and sea level lowering. *Journal of Geology*, **70**: 206–214.

Doody, J.P. (1991) Sea Defence and Nature Conservation: Threat or Opportunity? Paper presented to the Institution of Water and Environmental Management, Winter meeting.

Dubois, R.N. (1977) Predicting beach erosion as a function of rising water level. *Journal of Geology*, **85**: 470–476.

Economic Planning Unit (Malaysia) (1985) *National Coastal Erosion Study*. Government of Malaysia, Kuala Lumpur.

Edelman, T. (1972) Dune erosion during storm conditions. In: *Proceedings of the 13th Conference on Coastal Engineering*. American Society of Coastal Engineers, pp. 1305–1312.

Edwards, A.J. (1989) *The Implications of Sea-Level Rise for the Republic of Maldives*. Centre for Tropical Coastal Management Studies, Newcastle upon Tyne.

Eid, E.M. and Hulsbergen, C.H. (1991) Sea level rise and coastal zone management. In: J. Jaeger and H.L. Ferguson (eds.) *Climate Change: Science, Impacts and Policy*. Cambridge University Press, Cambridge, pp. 301–310.

Ellison, J.C. and Stoddart, D.R. (1991) Mangrove ecosystem collapse during predicted sea-level rise: Holocene analogues and implications. *Journal of Coastal Research*, **7**(1): 151–165.

El Sayed, M.K. (1991) Implications of relative sea level rise on Alexandria. In: R. Frassetto (ed.) *Impacts of Sea Level Rise on Cities and Regions. Proceedings of the First International Conference 'Cities on Water'*. Marsilio Editori, Venice, pp. 183–189.

Emanuel, K.A. (1987) The dependence of hurricane intensity on climate. *Nature*, **326**: 483–485.

Emery, K.O. and Aubrey, D.G. (1991) *Sea Levels, Land Levels and Tide Gauges*. Springer-Verlag, Berlin.

Everts, C.H. (1985) Sea level rise effects on shoreline position. *Journal of Waterway, Port, Coastal and Ocean Engineering*, **111**: 985–999.

Ewing, J.R. (1962) Pleistocene ice volumes and sea level lowering. *Journal of Geology*, **70**: 206–214.

Fabbri, P. (1985) Coastline variations in the Po delta since 2500 BP. *Zeitschrift für Geomorphologie, Supplementband*, **57**: 155–167.

Fairbridge, R.W. (1947) A contemporary eustatic rise in sea level. *Geographical Journal*, **109**: 157.

Fairbridge, R.W. (1961) Eustatic changes in sea level. *Physics and Chemistry of the Earth*, **4**: 99–185.

Fairbridge, R.W. (1966) Mean sea level changes. In: *Encyclopaedia of Oceanography*, Van Nostrand Reinhold, New York, pp. 479–485.

Fairbridge, R.W. (1987) The spectra of sea level in a Holocene time frame. In: M.R. Rampino, J.E. Sanders, W.S. Newman and L.K. Königsson (eds.) *Climate: History, Periodicity and Predictability*. Van Nostrand Reinhold, New York, pp. 127–142.

Fairbridge, R.W. and Krebs, O.A. (1962) Sea level and the Southern Oscillation. *Geophysical Journal*, **6**: 532–545.

Fisher, J.J. (1984) Regional long-term and localized short-term coastal environmental geomorphology inventories. In: J.E. Costa, and P.J. Fleischer (eds.) *Developments and Applications of Geomorphology*. Springer-Verlag, Berlin, pp. 68–96.

Frassetto, R. (ed.) (1991) *Impacts of Sea Level Rise on Cities and Regions. Proceedings of the First International Conference 'Cities on Water'*. Marsilio Editori, Venice.

Gagliano, S.M., Meyer-Arendt, K.J. and Wicker K.M. (1981) Land loss in the Mississippi deltaic plain. *Transactions of the 31st Annual Meeting of the Gulf Coast Association Geological Society*, pp. 293–300.

Gambolati, G., Ricceri, G., Bertoni, W., Brighenti, G. and Viullermin, E. (1991) Mathematical simulation of the subsidence of Ravenna. *Water Resources Research*, **27**(11): 2899–2918.

Gatto, P. and Carbognin, L. (1981) The lagoon of Venice, natural environmental trend and man-induced modifications. *Hydrological Sciences Bulletin*, **26**: 379–395.

Gibb, J.G. and Aburn, J.H. (1986). *Shoreline Fluctuations and an Assessment of a Coastal Hazard Zone along Pauanui Beach, Eastern Coromandel Peninsula, New Zealand. Water and Soil Technical Publication no. 27*, Ministry of Works and Development, Wellington, New Zealand.

Gibbs, M. (1984) Economic analysis of sea level rise: methods and results. In: M.G. Barth and J.G. Titus (eds.) *Greenhouse Effect and Sea Level Rise*. Van Nostrand Reinhold, New York, pp. 215–250.

Gibbs, M. (1986) Planning for a sea level rise under uncertainty: a case study of Charleston, South Carolina. In: J.G. Titus (ed.) *Effects of Changes in Stratospheric Ozone and Global Climate*, vol. 4. US Environment Protection Agency, Washington, DC, pp. 57–72.

Gilbert, S and Horner, R. (1984) *The Thames Barrier*. Thomas Telford, London.

Goemans, T. (1986) The sea also rises: the ongoing dialogue of the Dutch with the sea. In: J.G. Titus (ed.) *Effects of Changes in Stratospheric Ozone and Global Climate*, vol. 4. US Environment Protection Agency, Washington, DC, pp. 47–56.

Goldsmith, P. and Hieber, S. (1991) Space techniques support the monitoring of sea level. In: R. Frassetto (ed.) *Impacts of Sea Level Rise on Cities and Regions. Proceedings of the First International Conference 'Cities on Water'*. Marsilio Editori, Venice, pp. 220–226.

Gornitz, V. and Kanciruk, P. (1989) Assessment of global coastal hazards from sea level rise. In: *Proceedings of the 6th Symposium of Coastal and Ocean Management*. American Society of Coastal Engineers, pp. 1345–1359.

Gornitz, V. and Lebedeff, S. (1987) Global sea level changes during the past century. In: D. Nummedal, O.H. Pilley and J.D. Howard (eds.) *Sea-Level Change and Coastal Evolution*. Special publication no. 41, Society of Economic Paleontologists and Mineralogists.

Goudie, A. (1986) *Environmental Change*. Clarendon Press, Oxford.

Guilcher, A. (1965) *Précis d'Hydrologie: Marine et Continentale*. Masson, Paris.

Guilcher, A. (1981) Shoreline changes in salt marshes and mangrove swamps (mangals) within the past century. In: E.C.F. Bird and K. Koike (eds.) *Coastal Dynamics and Scientific Sites*. Komazawa University, Tokyo, pp. 31–53.

Guilcher, A. (1988) *Coral Reef Geomorphology*, Wiley, Chichester.

Gutenberg, B. (1941) Changes in sea level, postglacial uplift, and the mobility of the Earth's interior. *Bulletin of the Geological Society of America*, **52**: 721–772.

Hands, E.B. (1983) The Great Lakes as a test model for profile response to sea level changes. In: P.D. Komar (ed.) *Handbook of Coastal Processes and Erosion*. CRC Press, Boca Raton, Florida, pp. 176–189.

Hansen, J.E., Lacis, A.A., Rind, D.H. and Russell, G.L. (1984) Climate sensitivity to increasing greenhouse gases, In: M.G. Barth and J.G. Titus (eds.) *Greenhouse Effect and Sea Level Rise*. Van Nostrand Reinhold, New York, pp. 57–77.

Hekstra, G.P. (1986) Will climatic changes flood the Netherlands? Effects on agriculture, land use and well-being. *Ambio*, **17**: 316–326.

Henderson-Sellers, A. and Blong, R. (1989) *Greenhouse Effect: Living in a Warmer Australia*. New South Wales University Press, Sydney, Australia.

Hendry, M. (1988) Implications of future sea level rise for Caribbean shorelines. In: *Proceedings of the International Symposium on Theoretical and Applied Aspects of Coastal and Shelf Evolution*, Amsterdam, pp. 37–39.

Herd, D.G., Yourd, T.L., Hansjurgen, C., Person, W.J. and Mendoza, C. (1981) The great Tumaco, Columbia, earthquake of 12 December 1979. *Science*, **211**: 441–445.

Hoffman, J.S. (1984) Estimates of future sea level rise. In: M.G. Barth and J.G. Titus (eds.) *Greenhouse Effect and Sea Level Rise*. Van Nostrand Reinhold, New York, pp. 79–103.

Hofstede, J.L.A. (1991) Sea level rise in the Inner German Bight (Germany) since AD 600 and its implications upon tidal flats geomorphology. In: H. Brückner and U. Radtke (eds.) *Von der Nordsee bis zum Indischen Ozean*. Franz Steiner Verlag, Stuttgart, pp. 11–27.

Hopley, D. (1982) *The Geomorphology of the Great Barrier Reef*. Wiley Interscience, New York.

Hopley, D. (1989) Coral reefs: zonation, zonality and gradients. In: E.C.F. Bird and D. Kelletat (eds.) *Zonality of Coastal Geomorphology and Ecology*. Essener Geographischen Arbeiten, vol. 18, pp. 79–123.

Hopley, D. and Kinsey, D.W. (1988) The effects of a rapid short-term sea-level rise on the Great Barrier Reef. In: G.I Pearman (ed.) *Greenhouse: Planning for Climatic Change*. CSIRO Division of Atmospheric Research, Melbourne, Australia, pp. 189–201.

Houghton, J.T., Jenkins, G.J. and Ephraums, J.J. (eds.) (1990) *Scientific Assessment of Climate Change*. Cambridge University Press, Cambridge.

Hsu, T.L. (1985) Taiwan. In: E.C.F. Bird and M.L. Schwartz (eds.) *The World's Coastline*. Van Nostrand Reinhold, New York, pp. 829–831.

Hubbard, D.K. (1985) What do we mean by reef growth? In: *Proceedings of the 5th International Coral Reef Congress*, vol. 6, pp. 433–438.

Hulsbergen, C.H. and Schröder, P. (1989) *Republic of Maldives: implications of sea level rise*. Ministry of Economic Affairs, The Netherlands.

Jeftic, L., Milliman, J.D. and Sestini, G. (eds.) (1992) *Environmental and Societal Implications of Climatic Change and Sea Level Rise in the Mediterranean*. Arnold, London.

Jelgersma, S. (1988) A future sea level rise: its impacts on coastal lowlands. In: *Atlas of Urban Geology*, vol. 1, *Geology and Urban Development*. ESCAP, United Nations Bangkok, pp. 61–81.

Jennings, J.N. (1971) Sea level change and land links. In: D.J. Mulvaney and J. Golson (eds.) *Aboriginal Man and Environment in Australia*. Australian National University Press, Canberra, pp. 1–13.

Jones, M.D.H. and Henderson-Sellers, A. (1990) History of the greenhouse effect. *Progress in Physical Geography*, **14**: 1–18.

Jones, T. (1991) A methodological framework for analysing the socio-economic impacts of sea level rise. In: R. Frassetto (ed.) *Impacts of Sea Level Rise on Cities and Regions. Proceedings of the First International Conference 'Cities on Water'*. Marsilio Editori, Venice, pp. 205–216.

Juliano, R.O, Anderson, J. and Librero, A.R. (1982) Philippines: perceptions, human settlements and resource use in the coastal zone. In: Chandra Soysa, Chia Lin Sien and W.L. Collier (eds.) *Man, Land and Sea*. Agricultural Development Council, Bangkok, pp. 219–240.

Kaplin, P.A. (1989) Shoreline evolution during the 20th century. In: A. Ayale-Castañeres, W. Wooster and A. Yañez-Aranciba (eds.) *Oceanography 1988*. National University Press, Mexico.

Kearney, M.S. and Stevenson, J.C. (1991) Island land loss and marsh vertical accretion rate evidence for historical sea-level changes in Chesapeake Bay. *Journal of Coastal Research*, **7**: 403–415.

Kidson, C. (1986) Sea level changes in the Holocene. In: O. Van de Plassche (ed.) *Sea-Level Research: A Manual for the Collection and Evaluation of Data*. Geo Books, Norwich, pp. 27–64.

Kidson, C and Heyworth, A. (1979) Sea 'level'. In: *Proceedings of the 1978 International Symposium on Coastal Evolution in the Holocene*, São Paulo, Brazil, pp. 1–27.

Kinsey, D.W. and Davies, P.J. (1979) Effects of elevated nitrogen and phosphorus on coral reef growth. *Limnology and Oceanography*, **24**: 935–940.

Knighton, A.D., Mills, K. and Woodroffe, C.D. (1991) Tidal-creek extension and saltwater intrusion in northern Australia. *Geology*, **19**: 831–834.

Kolb, C.R. and Van Lopik, J.R. (1966) Depositional environments of the Mississippi deltaic plain. In M.L. Shirley (ed.) *Deltas in their Geologic Framework*. Houston Geological Society, pp. 18–63.

Komar, P.D. (1986) The 1982–83 El Niño and erosion on the coast of Oregon. *Shore and Beach*, **64**: 3–12.

Komar, P.D. and Enfield, D.B. (1987) Short-term sea level changes and coastal erosion. In: D. Nummedal, O.E. Pilkey and J.D. Howard (eds.) *Sea Level Fluctuations and Coastal Evolution*. Tulsa special publication no. 41, pp. 17–28.

Kraft, J.C., Biggs, R.B. and Halsey, S.D. (1973) Morphology and vertical sedimentary sequence models in Holocene transgressive barrier systems. In: D.R. Coates (ed.) *Coastal Geomorphology*. State University of New York, Binghampton, pp. 321–354.

Krysanova, V., Meiner, A., Roosaare, J. and Vasilyer, A. (1989) Simulation modelling of coastal water pollution from an agricultural watershed. *Ecological Modelling*, **49**: 7–29.

Kunstadter, P., Sabasri. S. and Bird, E.C.F. (1986) *Man in the Mangroves*. United Nations University, Tokyo, Japan.

Leatherman, S.P. (ed.) (1981) *Overwash Processes*. Dowden, Hutchinson and Ross, Stroudsburg, Pennsylvania.

Leatherman, S.P. (1984) Coastal geomorphic response to sea level rise: Galveston Bay, Texas. In: M.G. Barth and J.G. Titus (eds.) *Greenhouse Effect and Sea Level Rise*. Van Nostrand Reinhold, New York, pp. 151–178.

Leatherman, S.P. (1990) Modelling shore response to sea-level rise on sedimentary coasts. *Progress in Physical Geography*, **14**(4): 447–464.

Leontiev, O.K. and Veliev, K.A. (1990) Western coasts of the Caspian Sea. In: O.K. Leontiev and K.A. Veliev (eds.) *Transgressive Coasts: Studies and Economic Development*. Moscow State University, Moscow, pp. 1–71.

Lewis, J.R. (1964) *The Ecology of Rocky Shores*. English Universities Press, London.

Lisitzin, E. (1974) *Sea Level Changes*. Elsevier, Amsterdam.

Louisse, C.J. and Van der Meulen, F. (1991) Future coastal defence in The Netherlands: strategies for protection and sustainable development. *Journal of Coastal Research*, **7**(4): 1027–1041.

May, V.J. and Heeps, C. (1985) The nature and rates of change on chalk coastlines. *Zeitschrift für Geomorphologie, Supplementband*, **57**: 81–94.

Mayuga, M.N. and Allen, D.R. (1969) *Subsidence in the Wilmington Oil Field, Long Beach, California, USA*. Association Internationale d'Hydrologie Scientifique, Actes du Colloque de Tokyo, Publication no. 88, pp. 66–79.

Mehta, A.J. and Cushman, R.M. (eds.) (1989) *Proceedings of the Workshop on Sea Level Rise and Coastal Processes*. Department of the Environment, Florida.

Mercer, J.H. (1978) West Antarctic ice sheet and CO_2 greenhouse effect: a threat of disaster. *Nature*, **271**: 321–325.

Milliman, J.D. (ed.) (1988) *Sea-Level Rise and Coastal Subsidence: Problems and Strategies*. Papers presented to the SCOPE Workshop on Subsiding Coasts, Bangkok, November 1988.

Milliman, J.D., Broadus, J.M. and Gable, F. (1989) Environmental and economic implications of rising sea level and subsiding deltas: the Nile and Bengal examples. *Ambio*, **18**: 340–345.

Mimura, M., Isobe, M. and Nadaoka, K. (1991) Impacts of sea level rise and a framework for their assessment. In: *Proceedings of the International Conference on Climate Impacts on the Environment and Society*, Ibaraki, Japan, pp. B45–50.

Miossec, A. (1991) *Defense des Côtes et Protection du Littoral*. Université de Nantes, France.

Misdorp, R., Steyaert, F., De Ronde, J. and Hallie, F. (1989) Monitoring the western part of the Dutch Wadden Sea – sea level and morphology. *Helgoländer Meeresuntersuchungen*, **43**: 333–345.

Mörner, N.A. (1976) Eustacy and geoid changes. *Journal of Geology*, **84**: 123–151.

Mörner, N.A. (1985) Models of global sea level changes. In: M.J. Tooley (ed.) *Sea Level Changes*. Blackwell, Oxford.

Motamed, A. (1991) *Cause de la Fluctuation Recente de Niveau de la Mer Caspienne*. IGCP Project 274 annual report, pp. 126–127.

Mulrennan, M.E. (1992) *Coastal Management: Challenges and Changes in the Torres Strait Islands*. Discussion paper no. 5, North Australian Research Unit, Darwin.

Neumann, A.C. and Macintyre, I. (1985) Reef response to a sea level rise: keep-up, catch-up or give-up. In: *Proceedings of the 5th International Coral Reef Congress*, vol. 3, pp. 105–110.

Nunn, P. (1990) Recent coastline changes and their implications for future changes in the Cook Islands, Fiji, Kiribati, the Solomon Islands, Tonga, Tuvalu, Vanuata and Western Samoa. In: J.C. Pernetta and P.J. Hughes (eds.) *Potential Impacts of Climatic Change in the Pacific*. UNEP Regional Seas Reports and Studies, p. 128.

Nurse, L.A. (1992) Predicted sea level rise in the Wider Caribbean: likely consequences and response options. Paper presented to the International Meeting on Regional Programmes and Environmental Protection, Genoa, Italy, February 1992.

Olson, J.S. (1958) Lake Michigan dune development, 3: lake level, beach, dune. *Journal of Geology*, **66**: 473–483.

Orlova, G. and Zenkovich, V.P. (1974) Erosion on the shores of the Nile delta. *Geoforum*, **18**: 68–72.

Orson, R., Panagetou, W. and Leatherman, S.P. (1985) Response of tidal salt marshes of the United States Atlantic and Gulf coasts to rising sea levels. *Journal of Coastal Research*, **1**: 29–37.

Park, R.A., Armentano, T.V. and Cloonan, C.L. (1986) Predicting the effects of sea level rise in coastal wetlands. In: J.G. Titus (ed.) *Effects of Changes in Stratospheric Ozone and Global Climate*, vol. 4. US Environment Protection Agency, Washington, DC, pp. 129–152.

Parry, M., Magalhaes, A.R. and Nguyen Huu Ninh (eds.) (1991) *The Potential Socio-economic Effects of Climatic Change*. United Nations Environment Programme.

Paw, J.M. and Chua Thia-Eng. (1991) Climatic change and sea level rise: implications on coastal areas utilization and management in south-east Asia. *Ocean and Shoreline Management*, **15**: 204–232.

Pearman, G.I. (ed.) (1988) *Greenhouse: Planning for Climatic Change*. CSIRO Division of Atmospheric Research, Melbourne, Australia.

Peerbolte, E.B., De Ronde, J.G. and Baarse, G. (1991) *Impact of Sea Level Rise on Society: A Case Study for The Netherlands*. UNEP and Government of The Netherlands.

Pernetta, J.C. and Elder, D.L. (1990) *Climate, Sea Level Rise and the Coastal Zone: Management and Planning for Global Changes*. International Union for the Conservation of Nature, Gland, Switzerland.

Pernetta, J.C. and Hughes, P.J. (1992) *Potential Impacts of Climatic Changes in the Pacific*. UNEP Regional Seas Reports and Studies.

Pernetta, J.C. and Osborne, P.L. (1988) Deltaic floodplains: the Fly River and the mangroves of the Gulf of Papua. In: *Potential Impacts of Greenhouse Gas Generation on Climatic Change and Projected Sea Level Rise on Pacific Islands*. UNEP, Split, pp. 94–111.

Pernetta, J.C. and Sestini, G. (1989) *The Maldives and the Impact of Expected Climatic Changes*. UNEP Regional Seas Reports and Studies, vol. 104. UNEP, Nairobi.

Pethick, J.S. (1981) Long-term accretion rates on tidal salt marshes. *Journal of Sedimentary Petrology*, **51**: 571–577.

Pethick, J.S. (1984) *An Introduction to Coastal Geomorphology*. Arnold, London.

Phillips, J.D. (1986) Coastal submergence and marsh fringe erosion. *Journal of Coastal Research*, **2**: 427–436.

Pirazzoli, P.A. (1980) Palaeogeographic interpretation of a peat layer at Torson di Sotto (Lagoon of Venice, Italy). *Eiszeitalter und Gegenwart*, **30**: 253–259.

Pirazzoli, P.A. (1983) Flooding ('aqua alta') in Venice (Italy): a worsening phenomenon. In: E.C.F. Bird and P. Fabbri (eds.) *Coastal Problems of the Mediterranean Sea*. University of Bologna, Italy, pp. 23–31.

Pirazzoli, P.A. (1986a) Secular trends of relative sea-level changes indicated by tide-gauge records. *Journal of Coastal Research*, Special Issue **1**: 1–26.

Pirazzoli, P.A. (1986b) Marine notches. In: O. Van de Plassche (ed.) *Sea-Level Research: A Manual for the Collection and Evaluation of Data*. Geo Books, Norwich, pp. 361–499.

Pirazzoli, P.A. (1987) Recent sea level changes and related engineering problems in the Lagoon of Venice (Italy). *Progress in Oceanography*, **18**: 323–346.

Pirazzoli, P.A. (1989) Recent sea-level changes in the North Atlantic. In: D.B. Scott et al. (eds.) *Late Quaternary Sea-level Correlations and Applications*. Kluwer Academic Publishers, Dordrecht, pp. 153–167.

Pirazzoli, P.A. (1991a) *World Atlas of Holocene Sea-Level Changes*. Elsevier Oceanography Series, vol. 58.

Pirazzoli, P.A. (1991b) Possible defences against a sea level rise in the Venice area, Italy. *Journal of Coastal Research*, **7**: 231–248.

Pirazzoli, P.A. and Montaggioni, L.F. (1986) Late Holocene sea-level in the northwest Tuamotu Islands, French Polynesia. *Quaternary Research*, **25**: 350–368.

Platt, H.R., Warrick, R.A. and Wigley, T.M.L. (1983) *Cities on the Beach: Management Issues of Developed Coastal Barriers*. Research paper no. 224, Department of Geography, University of Chicago.

Psuty, N.P. (1986) Impacts of impending sea-level rise scenarios: the New Jersey barrier island responses. *Bulletin of the New Jersey Academy of Science*, **31**: 29–36.

Psuty, N.P. (1992) Estuaries: challenges for coastal management. In P. Fabbri (ed.) *Ocean Management and Global Change*. Elsevier Applied Science, Amsterdam, pp. 502–520.

Pugh, D.T. (1987a) *Tides, Surges and Sea-Level. A Handbook for Engineers and Scientists*. Wiley, Chichester.

Pugh, D.T. (1987b) The global sea level observing system. *Hydrographic Journal*, **45**: 5–8.

Pugh, D.T. (1991) Cities on water: sea level measurements and planning. In: R. Frassetto (ed.) *Impacts of Sea Level Rise on Cities and Regions. Proceedings of the First International Conference 'Cities on Water'*. Marsilio Editori, Venice, pp. 217–219.

Ramnarong, V. and Buapeng, S. (1991) Mitigation of groundwater crisis and land subsidence in Bangkok. *Journal of Thai Geoscience*, **1**: 125–137.

Ranwell, D.S. (1964) Spartina marshes in Southern England. *Journal of Ecology*, **52**: 79–105.

Ranwell, D.S. (1972) *Ecology of Salt Marshes and Sand Dunes*. London.

Raper, S.C.B., Warrick, R.A. and Wigley. T.M.L. (1988) *Global Sea Level Rise: Past and Future*. Paper presented to the SCOPE Workshop on Subsiding Coasts, Bangkok, November 1988.

Redfield, A.C. (1972) Development of a New England salt marsh. *Ecological Monographs*, **42**: 201–237.

Reed, D.J. (1990) The impact of sea-level rise on coastal salt marshes. *Progress in Physical Geography*, **14 (4)**: 465–481.

Revelle, R.R. (1983) Probable future changes in sea level resulting from increased carbon dioxide. In: *Changing Climate: Report of the Carbon Dioxide Assessment Committee*. National Academy Press, Washington, DC, pp. 433–488.

Rosen, P.S. (1978) A regional test of the Bruun Rule in shoreline erosion. *Marine Geology*, **28**: M7–M16.

Ruddle, K. and Johannes, R.E. (eds.) (1985) *The Traditional Knowledge and Management of Coastal Systems in Asia and the Pacific*. UNESCO Regional Office for Science and Technology in Southeast Asia, Jakarta.

Salinas, L.M., De Laune, R.D. and Patrick, W.H. (1986) Changes occurring along a rapidly submerging coastal area. Louisiana, USA. *Journal of Coastal Research*, **2(3)**: 269–284.

Schwartz, M.L. (1965) Laboratory study of sea level rise as a cause of erosion. *Journal of Geology*, **73**: 568–634.

Schwartz, M.L. (1967) The Bruun theory of sea level rise as a cause of shore erosion. *Journal of Geology*, **75**: 76–92.

Schwartz, M.L. (ed.) (1973) *Barrier Islands*. Dowden, Hutchinson and Ross, Stroudsburg, Pennsylvania.

Schwartz, M.L. and Bird, E.C.F. (eds.) (1990) Artificial beaches. *Journal of Coastal Research*, Special Issue **6**.

Scoffin, T.P. and Stoddart, D.R. (1979) The nature and significance of micro-atolls. *Philosophical Transactions, Royal Society, London, Series B*, **284**: 99–122.

Scoffin, T.P., Stearn, C.W., Boucher, D., Frydl, P., Hawkins, C.M., Hunter, I.G. and MacGeachy, J.G. (1980) Calcium carbonate budget of a fringing reef on the west coast of Barbados, Part II: erosion, sediments and internal structure. *Bulletin of Marine Science*, **30**: 475–508.

SCOR Working Group 89 (1991) The response of beaches to sea-level changes: a review of predictive models. *Journal of Coastal Research*, **7**(3): 895–921.

Sestini, G. (1991) Implications of climatic changes for the Nile delta. In: L. Jeftic, D.J. Milliman and G. Sestini (eds.) *Climate Change and the Mediterranean*. Arnold, London, pp. 535–600.

Sestini, G. (1992) Sea level rise in the Mediterranean region: likely consequence and response options. Paper presented at the Conference on Regional Programmes and Environmental Protection: An exchange of experiences between Mediterranean and Caribbean countries for the protection of their seas, Genoa, February 1992.

Sestini, G., Jeftic, L. and Milliman, J.D. (1990) *Implications of Expected Climatic Changes in the Mediterranean: An Overview*. UNEP Regional Seas Reports and Studies, vol. 103. UNEP, Nairobi.

Shennan, I. and Sproxton, I. (1991) Impacts of future sea level rise on the Tees estuary – an approach using a Geographical Information System. In: R. Frassetto (ed.) *Impacts of Sea Level Rise on Cities and Regions. Proceedings of the First International Conference 'Cities on Water'*. Marsilio Editori, Venice, pp. 119–134.

Shennan, I. and Tooley, M.J. (1987) Conspectus of fundamental and strategic research on sea level changes. In: M.J. Tooley (ed.) *Sea-Level Changes*. Blackwell, Oxford, pp. 371–390.

Shepard, F.P. and Wanless, H.R. (1971) *Our Changing Coastlines*. McGraw Hill, New York.

Smith, S.V. (1981) The Houtman Abrolhos islands: carbon metabolism of coral reefs at high latitudes. *Limnology and Oceanography*, **26**: 612–621.

Smith, S.V. and Kinsey, D.W. (1976) Calcium carbonate production, coral reef growth and sea-level change. *Science*, **194**: 937–939.

Somboon, J.R.P. (1990) Coastal geomorphic response to future sea-level rise and its implication for the low-lying areas of Bangkok Metropolis. *Tonan Ajia Kenkyu (Southeast Asian Studies)*, **28**: 154–170.

South Australian Coast Protection Board (1984) *Adelaide Coast Protection Strategy*. Department of Environment and Planning, Adelaide, Australia.

Stark, K.P. (1988) Designing for coastal structures in a greenhouse age. In: G.I Pearman (ed.) *Greenhouse: Planning for Climatic Change*. CSIRO Division of Atmospheric Research, Melbourne, Australia, pp. 161–176.

Steers, J.A. (1953) *The Sea Coast*. Collins, London.

Stevenson, J.C., Ward, L.G. and Kearney, M.S. (1986) Vertical accretion in marshes with varying rates of sea level rise. In: D.A. Wolfe (ed.) *Estuarine Variability*. Academic Press, New York, pp. 241–259.

Stewart, R.W., Kjerfve, B., Milliman, J. and Dwivedi, S.N. (1990) Relative sea level changes: a critical evaluation. *UNESCO Reports in Marine Science*, **54**.

Stoddart, D.R. (1971) Coral reefs and islands and catastrophic storms. In: J.A. Steers (ed.) *Applied Coastal Geomorphology*. Macmillan, London, pp. 155–197.

Stoddart, D.R. (1990) Coral reefs and islands and predicted sea-level rise. *Progress in Physical Geography*, **14**: 521–536.

Stoddart, D.R. and Reed, D.J. (1990) Sea level rise as a global geomorphic issue. *Progress in Physical Geography*, **14**: 441–536.

Stoddart, D.R. and Steers, J.A. (1977) The nature and origin of coral reef islands. In: O.A. Jones and R. Endean (eds.) *Biology and Geology of Coral Reefs*, vol. 4. Academic Press, New York, pp. 59–105.

Sullivan, M. and Hughes, P. (1989) *Sea Level Rise Report for Papua New Guinea: Effects of and Responses to a Rising Sea Level*. University of Papua New Guinea, Port Moresby.

Sunamura, T. (1988) Projection of future coastal cliff recession under sea level rise induced by the greenhouse effect: Nii-jima Island, Japan. *Transactions of the Japanese Geomorphological Union*, **9**(1): 17–33.

Sunamura, T. (1992) *Geomorphology of Rocky Coasts*. Wiley, Chichester.

Tegart, W.J.McG., Sheldon, G.W. and Griffiths, D.C. (1990) *Climatic Change: The IPCC Impacts Assessment*. Australian Government Publishing Service, Canberra.

Teh Tiong Sa (1992) A conceptual model for predicting shoreline response on the permatang coast of Malaysia. In: Voon Phin Keong and Tunku Shamsul Bahrin (eds.) *The View from Within: Geographical Essays on Malaysia and Southeast Asia*. University of Malaya, Kuala Lumpur, pp. 64–83.

Thom, B.G. and Roy, P.S. (1987) Sea-level rise and climate: lessons from the Holocene. In: G.I Pearman (ed.) *Greenhouse: Planning for Climatic Change*. CSIRO Division of Atmospheric Research, Melbourne, Australia, pp. 177–188.

Titus, J.G. (ed.) (1986a) *Effects of Changes in Stratospheric Ozone and Global Climate*. US Environment Protection Agency, Washington, DC (four volumes).

Titus, J.G. (1986b) The causes and effects of sea level rise. In J.G. Titus (ed.) *Effects of Changes in Stratospheric Ozone and Global Climate*, vol. 1: Overview. US Environment Protection Agency, Washington, DC, pp. 219–249.

Titus, J.G. (ed.) (1988) *Greenhouse Effect, Sea Level Rise and Coastal Wetlands*. US Environment Protection Agency, Washington, DC.

Titus, J.G. (1991) Greenhouse effect and coastal wetland policy: how Americans could abandon an area the size of Massachusetts at minimum cost. *Environmental Management*, **15**: 39–58.

Tooley, M.J. (ed.) (1985) *Sea Level Changes*. Blackwell, Oxford.

UNEP (1986) *Environmental Problems of the Marine and Coastal Area of Bangladesh: National Report*. UNEP Regional Seas Reports and Studies, vol. 75.

Valentin, H. (1952) Die Küsten der Erde. *Petermanns Geogr. Mitteilungen*, **246**.

Valentin, H. (1953) Present vertical movements of the British Isles. *Geographical Journal*, **119**: 299–305.

Valentin, H. (1971) Land loss at Holderness. In: J.A. Steers (ed.) *Applied Coastal Geomorphology*. Macmillan, London, pp. 116–137.

Van de Plassche, O. (ed.) (1986) *Sea-Level Research: A Manual for the Collection and Evaluation of Data*. Geo Books, Norwich.

Van der Muelen, F. (1990) European dunes: consequences of climatic changes and sea level rise. *Catena Supplement*, **18**: 51–60.

Vanderzee, M.P. (1988) Changes in saltmarsh vegetation as an early indicator of sea-level rise. In: G.I Pearman (ed.) *Greenhouse: Planning for Climatic Change*. CSIRO

Division of Atmospheric Research, Melbourne, Australia, pp. 147–160.

Vellinga, P. (1988) Sea level rise, consequences and policies. In: P.C. Schröder (ed.) *Sea Level Rise: A Selective Retrospection*. Delft Hydraulics, The Netherlands.

Viles, A.H. (1989) The greenhouse effect, sea level rise and coastal geomorphology. *Progress in Physical Geography*, **13**: 422–461.

Vongvisessomjai, S. (1990) Coastal erosion in Thailand. *Geographical Journal of Thailand*, **15**: 321–337.

Walker, D. (1972) *Bridge and Barrier: The Natural and Cultural History of Torres Strait*. ANU Press, Canberra.

Walker, H.J. (1988) *Artificial Structures and Shorelines*. Kluwer Academic Publications, New York.

Wanless, H.R. (1982) Sea level is rising – so what? *Journal of Sedimentary Petrology*, **52**: 1051–1054.

Warwick, R. and Farmer, G. (1990) The greenhouse effect, climatic change and rising sea level: implications for development. *Transactions, Institute of British Geographers*, **15**: 5–20.

Wigley, T.M.L. and Raper, S.C.B. (1987) Thermal expansion of sea water associated with global warming. *Nature*, **330**: 127–131.

Williams, E.H. and Williams, L.B. (1990) Coral reef bleaching alert. *Nature*, **346**: 225.

Wind, H.G. (ed.) (1987) *Impact of Sea Level Rise on Society*. Balkema, Rotterdam, The Netherlands.

Wong Poh Poh (1985) Artificial coastlines: the example of Singapore. *Zeitschrift für Geomorphologie, Supplementband*, **57**: 175–192.

Woodroffe, C.D. (1988) Changing mangrove and wetland habitats over the last 8,000 years, Northern Australia and Southeast Asia. In: D. Wade-Marshall and P. Loveday (eds.) *Floodplains Research, Northern Australia: Progress and Prospects*, vol. 2. Australian National University, Canberra, pp. 1–33.

Woodroffe, C.D. (1989) Salt water intrusion into groundwater: an assessment of effects on small island states. Paper delivered to Small Island States Conference on Sea Level Rise, Male, November 1989.

Woodroffe, C.D. (1990) The impact of sea-level rise on mangrove shorelines. *Progress in Physical Geography*, **14**(4): 483–520.

Woodroffe, C.D. and McLean, R. (1990) Microatolls and recent sea level change on coral reefs. *Nature*, **344**: 531-534.

Woodroffe, C.D., Thom, B.G. and Chappell, J. (1985) Development of widespread mangrove swamps in mid-Holocene times in northern Australia. *Nature*, **317**: 711–713.

Woodworth, P.L. (1987) Trends in UK mean sea level. *Marine Geodesy*, **11**: 57–87.

Woodworth, P.L. (1991) The Permanent Service for mean sea level and the Global Sea Level Observing System. *Journal of Coastal Research*, **7**: 699–719.

Wroblewski, J.S. and Hoffmann, E.E. (1989) US interdisciplinary modelling studies of coastal-offshore exchange processes: past and future. *Progress in Oceanography*, **23**: 65–99.

Wyrtki, K. (1990) Sea level rise: the facts and the future. *Pacific Science*, **44**: 1–16.

Yap, H.T. (1989) Implications of expected climatic changes on natural coastal ecosystems in the East Asian Seas region. In: Chou Loke Ming (ed.) *Task Team Meeting on Implications of Climatic Change in the East Asian Seas Region*. University of Singapore, pp. 128–148.

Yohe, G.W. (1990) Towards an analysis of policy, timing, and the value of information in the face of uncertain greenhouse-induced sea level rise. In: J.G. Titus (ed.) *Changing Climate and the Oceans* vol. 1. Environmental Protection Agency, Washington, DC, pp. 353–372.

Zanda, L. (1991) The case of Venice. In: R. Frassetto (ed.) *Impacts of Sea Level Rise on Cities and Regions. Proceedings of the First International Conference 'Cities on Water'*. Marsilio Editori, Venice, pp. 51–59.

Zunica, M. (1990) Beach behaviour and defences along the Lido di Jesolo, Gulf of Venice, Italy. *Journal of Coastal Research*, **6**: 709–719.

Author Index

Geographical Index

Subject Index

Abrasion, 45
Accretion, 82, 85, 89, 134
  in salt marshes, 73, 75, 78
  vertical, 78
Accretionary growths, 45
Acropora, 93
Aggradation, 78
Algal cultures, 150
  encrustations, 45
Americam National Flood Insurance
    Program, 136, 148
American Sea Level Datum, 9
Aqua alta (Venice), 13, 113
Aquaculture, 81, 133, 135, 138, 141
Aquifer, 19, 61
Artificial coasts, 98–101
Artificial entrance, 64
Artificial intertidal shoals, 138
Artificial islands, 139, 150
Artificial nourishment, *see* Beach
    nourishment, artificial
Artificial structures, 37, 40, 136
Artificial submerged reefs, 146
Atolls, 89, 90, 96, 135
  drowned, 94
  micro, 95
  tilted, 96
Australian aborigines, 128
Australian Resources Information
    System, 120

Backshore zone, 5
Banks, 81
Bar formations, 52
Barene, 62, 76

Barrages, 114
Barrier (Barrier island), 50, 54, 55,
    58–62, 64, 71, 104, 105, 113, 116,
    121, 126, 127
  formation, 62
  prograded, 50
  reefs, *see* Reefs
  transgressive, 50, 56
Basal notch, 46
Basal talus, 46, 47
Beach accretion, advance, progradation,
    48, 50, 54, 55
Beach, beaches, 42, 48–60, 79, 100
  artificial, 100, 107
  behaviour, 150
  drift-dominated, 49, 50
  huts, 126
  quarrying, 52
  raised, 40
  resorts, 99, 100, 136
  ridges, 50, 51, 55, 58, 148
  rock, 97
  swash-dominated, 49
  system, 51
Beach erosion, causes, 53
  depletion, 48, 51, 52, 54, 56, 57, 58,
    99–101, 104, 106, 111, 113, 125,
    127, 136
Beach nourishment, artificial, 60, 100,
    105, 108, 113, 136, 137, 147, 149
Beach-ridge plain, 50, 54, 72, 125
Biodiversity, 37
Biological zonations, 33–35
Bluffs, 44, 45
Boat ramps, 126
Boatsheds, 126